REINVENTING THE CHICKEN COOP

REINVENTING THE CHICKEN COOP

14 Original Designs
with
Step-by-Step
Building
Instructions

KEVIN McELROY &
MATTHEW WOLPE

PHOTOGRAPHY BY ERIN KUNKEL

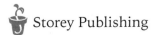
Storey Publishing

The mission of Storey Publishing is to serve our customers by publishing practical information that encourages personal independence in harmony with the environment.

Edited by Philip Schmidt, Sarah Guare, and Deborah Balmuth
Art direction and book design by Alethea Morrison
Text production by Jennifer Jepson Smith
Photography by © Erin Kunkel
Illustrations by © Michael Gellatly

Storey Publishing
210 MASS MoCA Way
North Adams, MA 01247
www.storey.com

Printed in China by R.R. Donnelley
10 9 8 7 6 5 4 3 2 1

Library of Congress Cataloging-in-Publication Data on file

Acknowledgments

This book would not be possible without an incredible network of support from the following people:

Russell Brent and Nicholas Hodges, Leah Fessenden, Sabina Rasenow, Kaytlyn O'Connor, and Salvador Menjivar for housing our finished coops. Novella Carpenter for housing our coop and providing insight and guidance on all aspects of our project. Our contributors: Traci Fontyn, Adam Reineck, Yvonne Mouser, and Nicole Starnes Taylor. Maurice Cavness, Nelson Bellesheim, Michel Dedeo, Michael Seidel, and Darren McElroy for helping move some of our (ridiculously heavy) coops. Biofuel Oasis for spreading the word about Just Fine Design/Build. Julie Pavuk and Barbara Finnin of City Slicker Farms; we're so happy to have a long-standing relationship with their organization. Shannon Little for legal advice. Bill from ShedMover.com. Vale Larson Brasted and Dennis McElroy for invaluable design and construction advice. Erin Kunkel, our amazing photographer. Stephany Fillimon for technical writing advice. Joey Gottbrath and the studio staff at The Crucible for putting up with all of our construction and having a sense of humor about it. Deborah Balmuth, Sarah Guare, and Alethea Morrison for their ongoing support at Storey Publishing. Philip Schmidt for the careful and diligent editing. Tamar Beja, Audrey Hodtwalker, Hillary Kantmann, and Andrew Murphy for all-around chicken-keeping support. Andreas Stavropoulos for being a mentor and inspiration. Sean McElroy and Sean Tischler for help cutting the shipping container in half. Deborah Lebow, Alisa Lebow, Deena Lebow, Josh Wolpe, Beth Lebow, Sheri Kuehl, Jane Dawson, Kathy McElroy, Ryan & Jaime McElroy, and Chloe Sladden for editing help, ongoing support, encouragement, and love throughout this wacky project.

CONTENTS

PREFACE

We certainly did not set out to write a book on chicken coops. While debating whether to take on such an ambitious project, we were well aware of our shortcomings. We are not farmers, chicken experts, technical writers, architects, or ecologists, and we had never published anything of significance.

What we are, however, are makers — part of a strong current of people in the Bay Area who sketch and tinker, make messes and experiments in their basements and garages, and keep coming back for that addictive satisfaction of creating something useful. In 2008, Matt taught a collaborative design/build class in Oakland with City Slicker Farms. In the free class, community members learned basic carpentry skills while working on a project to take home. And the project? A chicken coop. To teach the class, Matt had to learn a lot about chicken coops fast.

Fast-forward a year. As close friends and fellow generalists, we formed Just Fine Design/Build in 2009 as a creative outlet and experimental business. When brainstorming a project to bring to Maker Faire (a convention of inventors, hackers, musicians, farmers, and artists), we decided to take another stab at chicken coop design, this time expanding upon the lessons learned from Matt's first round. We felt we had something to contribute to the body of chicken coops out there, to add to the discussion: something well-designed, contemporary, clean; a coop people wouldn't apologize for as an eyesore in their yard but instead present proudly as a statement of self-sufficiency; something well-built with thought and intention. Like an egg, out came Chick-in-a-Box (page 94).

The coop was featured on some design blogs and received an Editor's Choice Award at Maker Faire, but none of the thousands of festival patrons were willing to commission a custom coop as we hoped. People loved our coop, but we were competing against prefab kits sold on the Internet at a fraction of the cost. We were realizing the difficulty in making any money at custom design/build projects. We both worked other jobs to support our creative pursuits: Matt on staff at an arts nonprofit and teaching furniture-making classes, and Kevin as a winemaker.

Then one day, soon after Maker Faire, we were contacted by Storey Publishing. They had seen Chick-in-a-Box featured in a design publication and liked what they saw. Six months later, we submitted a proposal for the book you have in your hands, and we decided to take the chicken coop much more seriously. (Well, not too seriously!) It was an appealing project, sort of like building a bunch of very small houses, each with its own roofing, flooring, and framing system. It had the potential for the detail and craftsmanship of fine furniture, and the design vocabulary of architecture. It connected to bigger ideas of food production and sustainability, and it empowered others through the safe and effective use of tools.

With funding from the publisher, we could design custom coops and not have to worry about selling each one individually. Design inspiration came from many directions: materials, clients, sites, traditions, and our own dreams and ideas. The Pallet Coop (page 144) was designed and built for the legendary urban homesteader Novella Carpenter. We wanted to create a structure out of old shipping pallets, and she was the ideal candidate, as her farm is filled with recycled and castaway parts. The Cupe (page 168) was built for clients who live high in the hills of Berkeley, California, and features a roofline and primary colors mirroring those of their beautiful midcentury modern house. The Coopsicle (page 158) was designed to accommodate the slope of a steep hillside and also satisfied Matt's childhood desire to build a tree-house. The Container Coop (page 182) is built from a surplus 20-foot shipping container cut in half and looks right at home in an industrial part of West Oakland, California.

We are also pleased to present the work of other talented chicken coop designers. Nicole Starnes-Taylor and Traci Fontyn had already designed and built their backyard chicken coops and agreed to include their designs in our book. Adam Reineck and Yvonne Mouser had designed their SYM coop (page 106) for a competition but had not built it until we approached them.

With this collection of 14 coop designs, our hope is to expand the definition of what a chicken coop is or could be. Surely, building more time-tested coop structures would be sufficient, and we love the traditional coops out there. But the role of the chicken is expanding, and with it we see the nature of chicken shelter expanding, too. Chickens are no longer solely in the domain of the farm; now they are a fundamental component of DIY food production and food security in urban and suburban backyards. This makes coop design an ideal medium for experimentation: How can chicken coops better serve users in the contemporary world? How can they look and function differently? What kinds of materials can be used? Can chicken coops be treated like a piece of outdoor furniture? Can chicken coops serve multiple purposes in a well-functioning small urban farm? Our designs are an attempt to answer these types of questions.

During our year of frantically designing and building coops, we also had a whole lot of fun. What could be better than working with your friend, designing and building small structures, and having creative control of a project from start to finish? Unlike building a house, which could stretch on for years, a chicken coop is a project you can tackle in a matter of days — the perfectly scaled project to sink your teeth into and not get burned out on.

That year also brought us into contact with a cadre of characters: seasoned urban farmers, eccentric chicken enthusiasts, foodies, suburban families, landscape architects, structural engineers, industrial designers, permaculture practitioners, and single parents working to provide healthy food for their kids. All preconceived notions of who we would encounter when taking on this project were shattered. We found interest in keeping hens across wide swaths of the population. Interested folks were enthusiastic and certainly not shy about sharing their experiences and opinions about chickens. They either keep chickens currently or are planning to in the near future, or they have a friend with chickens, grew up on a farm, or want to tell us their grandmother's experience with chickens as a young girl.

We also found a new appreciation for the egg. It is nothing short of a miracle that such a perfect morsel of protein is produced by a healthy chicken daily — a feat of nature that is not to be taken for granted.

Now get out there and build your own coop! Follow our instructions word for word, or exercise your creative right and customize one of our coops to fit your lifestyle, your yard, and your personality.

Kevin McElroy and Matt Wolpe
Oakland, California

THE
BASICS

CHICKEN COOP ESSENTIALS

During the process of designing and building coops for this book, we became accustomed to quizzical looks when explaining our latest projects. There were plenty of positive responses, but just as often came the question, "Why would you want chickens?" It was an easy thing to answer: "For the eggs, of course." However, this wasn't the question we were hoping most to be asked: "Why would you want to *build* a chicken coop?"

What appeals to us about the form is that there's a set of design parameters for a well-designed coop, and beyond that it can really be anything. Also, most coops are informal structures and typically do not require building permits or messy bureaucracy. With coops, ordinary people can actually build something of architectural significance next to their home and be proud of the statement it makes: "I made this thing, and it brings me closer to the food I eat every day."

Along with our desire to explore the idea of what a chicken coop could be, like any thoughtful designers we wanted to be cautious and deliberate. A new idea isn't always a good one, and a classic, time-tested technique is not something to be dismissed. We wanted the design choices we made to be significant, not random, guided by the site and concept for each coop. We also wanted to experiment with conventional and unconventional building materials and to treat the chicken coop like the hybrid structure it is — part outbuilding, part outdoor furniture, part sculpture. Throughout the project we often joked with each other that our mission was to bring the coop from the backyard to the front yard.

Given the scope of what we were taking on, however, it was important to stick to our fundamental premise: To design really good coops for the chickens and their owners. An experimental design that looks incredible but functions poorly and doesn't pass the scrutiny of an experienced hen raiser would be a failure on our part. Therefore, we approached each coop project with the following design considerations in mind. The same guidelines and questions can help you when making modifications to your own projects.

GENERAL SPACE REQUIREMENTS

While there are many different breeds of chickens and, of course, different sizes, our rule of thumb is to provide roughly 8 to 10 square feet of space per chicken. This number is the coop and run space combined. Keep in mind that if you live in a cold climate where the chickens will be inside for a good portion of the year, you should increase your coop size so that it remains a healthy living environment.

MINIMUM SPACE REQUIREMENTS

BIRDS	AGE	OPEN HOUSING		CONFINED HOUSING	
		sq ft/bird	birds/sq m	sq ft/bird	birds/sq m
HEAVY	1–7 days	—	—	0.5	20
	1–8 weeks	1	10	2.5	4
	9–15 weeks (or age of slaughter)	2	5	5	2
	15–20 weeks	3	4	7.5	1.5
	21 weeks +	4	3	10	1
LIGHT	1 day–1 week	—	—	0.5	20
	1–11 weeks	1	10	2.5	4
	12–20 weeks	2	5	5	2
	21 weeks +	3	3	7.5	1.5
BANTAMS	1 day–1 week	—	—	0.3	30
	1–11 weeks	0.6	15	1.5	7
	12–20 weeks	1.5	7	3.5	3
	21 weeks +	2	5	5	2

This table was adapted from *The Chicken Health Handbook* by Gail Damerow. "Open housing" refers to a coop with a fenced range, while "confined housing" refers to a coop with an enclosed run (see page 14).

To encourage your chickens to spend more time outdoors, you can provide a covered area where they can be outside of their coop but somewhat protected from the elements. More time outside results in healthier chickens and a cleaner coop. As always, the more space you can offer the better. Free-range chickens who can roam for insects and have a safe place to sleep at night are the happiest.

COOP AND RUN SETUPS

For backyard chicken keeping, you can have a variety of different setups, each with its own advantages and disadvantages. Deciding which setup is right for you involves many considerations, such as the amount of available space, local zoning ordinances, weather, the number of chickens you'd like to keep, and how much time you want to spend on maintenance. Keep in mind that a coop alone — with no access to outdoor space and ground contact — is not ideal for backyard chicken keeping. There should always be a relationship between the enclosed coop and some yard space, and that arrangement can take many forms. Here are some classic examples.

DEEP BEDDING

Incorporating chicken raising into your household waste stream is not only a good sustainability practice; it can also offset your feed costs. To deal with the accumulation of chickens' droppings, some keepers simply clean out the coop and replace the bedding periodically. Others prefer the "deep bedding" technique: continually adding fresh straw or other bedding in generous amounts so that as the straw builds up the lower levels underneath begin to compost, providing rich soil that can be collected periodically and used in the garden.

The Cupe has been adapted to include an enclosed run.

The Container Coop was designed to have an enclosed run.

The Modern Mobile Coop-Tractor features a removable floor.

Coop and Enclosed Run (Confined Housing)

Many of the designs in this book combine a coop with an attached run, both secure from predators. The advantage here is that there's no need to let the chickens out of their coop in the morning and lock them up at night. They're always safe and can come and go between indoor and outdoor spaces as they please. This is probably the most low-maintenance of coop designs. If the chickens have enough food and water, they can even be left for several days while you are out of town (though you'll probably want to have someone check on them). The Container Coop, Pallet Coop, and Corner Coop are all built in this style.

One disadvantage to this style is that it requires more care and attention when building the coop and the run, because any weakness in the design can be exploited by crafty predators. Building a secure run also includes having a roof of some sort (which can be mesh) so that no birds can get in and predators can't jump in from trees. As with the fenced range option, another disadvantage is that the chickens will quickly eat through all of the vegetation in the fenced-in area and leave a generous amount of droppings. However, you can supplement their food supply with kitchen scraps and edible greens that might not make the cut into your salad (be sure to see What *Not* to Feed Chickens, on page 18).

Coop and Fenced Range (Open Housing)

If you like the idea of having "free range" chickens, and you're the type who enjoys the daily rituals of tending to them, a coop and fenced-in run setup is ideal. With this arrangement, the coop must be secure for the hens to roost at night, but the run simply has to contain the chickens and can be nothing more than a fence (at least 4 feet high) with no roof. This simplifies construction and allows for more design possibilities, though it does require letting the chickens out in the morning and locking them in at night. Also, the same issues of vegetation depletion and droppings accumulation noted with the coop and enclosed run apply here. Your hens will also be more vulnerable to predators if the run is built without a roof.

Chicken Tractor

The chicken tractor is a popular design that works like a very old-fashioned, but very natural, lawnmower. The coop is mobile and has no floor, allowing the chickens to forage, peck, and fertilize the ground directly underneath the coop. Once the immediate area is exhausted, the coop is moved to a new location with fresh ground.

We love the idea of chicken tractors but find that for small urban lots you might have to dedicate all of your yard to the chickens to provide enough fresh ground.

In a matter of just two or three hours, the chickens can eat all of the vegetation in a tractor. We think a better application of a tractor design is in a farm setting, where there is plenty of area for moving the tractor, allowing sufficient time for the vegetation to regenerate before the chickens get to it again. Another good option is a hybrid structure that can be both a tractor and a coop and run, if desired. Other drawbacks to tractor systems are that they tend to offer inadequate protection in cold weather, and when not set on level ground, they can create holes for predators to exploit.

Paddock Rotation System

The paddock system is a favorite of permaculture practitioners. While none of the coops in this book is designed for paddock rotation (due to the nature of the sites they were built on), some could be modified for this application. A paddock system involves dividing a yard into several sections — each with its own entrance to a stationary coop — and rotating the chickens from one paddock to another.

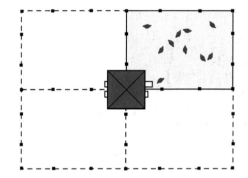

If you put small doors on each corner of the coop, you can rotate chickens between paddocks without having to move the coop.

As a variation of the tractor system, paddock rotation requires a decent amount of space to allow the vegetation to regenerate and the ground to become sanitized during the fallow periods, but it does not require moving the coop/tractor every few hours. The time needed for letting the paddocks rest varies on their size and the amount of chickens you have. An ideal setup might include four or six paddocks, each with 8 to 10 square feet of space per bird. There are also some two-paddock systems that work quite well, and the chickens are moved every six months. While accumulation of droppings is still an issue, this does give the resting paddock a chance to regenerate.

ROOSTS

Roosts, or roosting bars, are perches that chickens sleep on at night, similar to a tree branch if they were in the wild. A great deal of droppings will accumulate underneath the roosts, so having a well-designed, easily cleaned floor (or even a drawer) is a good idea. Chickens typically like to roost high up in the coop, but you can put your roosts pretty much anywhere. Allow for at least 8" of horizontal roosting space on the bar per bird and 12" of height. Also be sure to use wood that doesn't have sharp corners. A 2×2 with rounded edges works great. Making a roosting bar removable allows for easy cleaning and replacement. We like to cut a notch in each bar support so that a 2×2 can simply slip right into it.

A roosting bar rests in notched cleats of the Modern Log Cabin.

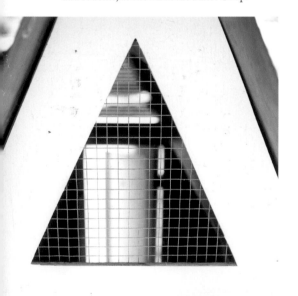

When installed correctly, wire mesh (hardware cloth) provides great protection from predators and rodents, as shown in the Pallet Coop.

A ventilation window in the A-Frame Coop

PROTECTION FROM PREDATORS AND RODENTS

In addition to comfortable shelter, a well-designed chicken coop provides a haven from predators, such as dogs, raccoons, foxes, and hawks. Keep any gaps in the coop itself to less than ½" to ensure protection. As we will discuss in the next chapter, wherever there is mesh, we suggest using ½" hardware cloth instead of traditional chicken wire. Hardware cloth has smaller openings, and it's more durable and easier to cut (during building) than chicken wire. (See page 26.)

While stapling the mesh creates a pretty strong connection, a better way to secure it is to sandwich it between a piece of 1×2 and the coop framing by screwing through the 1×2 and one of the openings in the mesh. This type of attachment prevents raccoons from prying off staples and using leverage to get into the coop. Similarly, use latches to close all doors and windows. Spring-loaded and lockable latches keep predators out, and if you live in a high-traffic urban environment you may consider a padlock to keep out the most industrious predators — humans!

Where the coop or run meets the ground is another important area for protection. Rather than ending the mesh at the ground, we suggest digging a trench and burying it 16" below the surface to keep out burrowing predators. While this is quite a bit of extra work up front, it can potentially save your chickens down the road and gives you peace of mind. If your coop is completely enclosed with a mesh floor, this step is unnecessary.

The primary rodents to keep out of your coop are rats. Though not necessarily a danger to the chickens themselves, rats do get in to eat the feed and wreak general havoc. Rats are industrious and can wiggle themselves into small openings, so take extra care with the design around vulnerable areas, such as door openings and roof joints.

VENTILATION

Ventilation is essential for a healthy chicken coop. It removes dust and moisture, prevents the buildup of ammonia from accumulated droppings, and reduces the likelihood of airborne diseases. We like to put mesh near the roof at both sides of the coop so that hot air can escape and a cross breeze can enter the interior. Keep in mind that during the winter reducing drafts is also important. This may mean covering one or more of the sides (depending on your location and the position of the coop) to provide a good balance between weather protection and ventilation.

Nesting boxes in the Pallet Coop

NESTING BOXES

Hens like to feel extra safe and secure when laying eggs, so a nesting box typically is a dark space with an enclosed feel. Without a nesting box, your hens may lay their eggs all over the place, so providing them with an ideal laying space ensures that you'll know where to find your eggs consistently. Also, each chicken does not need her own nest; we generally recommend one nest for every four to five chickens. The ideal dimensions for an individual laying box are 12" wide × 14" high × 12" deep, and it's usually best to arrange multiple boxes side by side, like little row houses. Since chickens leave a lot of droppings where they roost, it is best to avoid having the chickens roost in their nesting area.

A handy detail to include is a little lip on the front and back edges of the laying box. This will prevent eggs from rolling out and helps keep in nesting material (straw or shredded paper). Figuring out how to gather the eggs easily and quickly is a fun design challenge to take on. We have several different schemes for egg collection, but feel free to invent your own.

BEDDING MATERIAL
Bedding should be a nontoxic material that's kept dry and free of mold. Good options include chopped straw, shredded paper, and pine shavings (not hardwood shavings).

You will thank yourself down the road if you add a human-sized access door, as we did in the Corner Coop.

WHAT *NOT* TO FEED CHICKENS

Chickens will eat pretty much anything humans will eat and more, but here are some foods that don't belong in a chicken's diet:

- raw potato peels
- onions, garlic, fish (which can flavor the eggs)
- avocados
- spoiled or rotten food
- fried foods
- food containing caffeine or alcohol
- food high in fat, sugar, or sugar substitute
- hardwood and cedar shavings

HUMAN ACCESS

Perhaps the most important requirement for a chicken coop is that it's easy to clean. If cleaning is difficult, it increases the likelihood that cleaning will fall by the wayside. To prevent detritus from accumulating, we recommend smooth surfaces and a minimum of cracks in the floor of the coop. There should be at least one opening for humans to access the coop easily to scoop or sweep out soiled bedding.

In addition to a human access door, most of our coops have an additional door just for the chickens, commonly called a "pop hole." An added design feature would be an automated pop hole connected to a timer, which would allow for the chickens to be let out in the morning and locked in at night. We have seen some clever DIY designs for these.

CLIMATE CONTROL AND INSULATION

While most of the coops in this book were designed with the mild weather of northern California in mind, the Standard Coop (page 40) and A-Frame Coop (page 41) were built with stud walls spaced to accept standard house insulation. The other coops can be adapted to suit either warmer or colder climates. However, some coops are better candidates for weatherizing than others. These are the Corner Coop, Coopsicle, and Container Coop, and we have included tips for modifying them.

To keep your coop temperate (particularly from getting too hot) in more extreme climates, apply insulation to the roof and walls. One simple modification is to add rigid foam insulation board between the framing members, on the inside of the coop; just be sure to cover the insulation with some sort of interior siding to keep the chickens from pecking at it. Rigid foam insulation comes in a variety of thicknesses. Typically you want to match it to the thickness of your wall framing (for example, 1½"-thick insulation for 2×2 wall studs). Using aluminum or a light-colored roofing material can help reflect heat, as can painting the roof white and having plenty of shade from trees around the coop.

In very cold weather, you'll want insulation as well as an air and moisture barrier (such as housewrap) lining the walls. In a well-designed and -built structure, even in the dead of winter the chickens can remain at a comfortable temperature as long as they are not wet or subjected to an overly drafty environment. If your coop has a lot of mesh-covered openings, you may want to cover them for the winter, but be sure to leave enough open for adequate ventilation.

COOP-BUILDING BASICS

One of our goals for this book was to keep things simple, using ordinary shop tools and building with similar materials and repeatable processes as much as possible. So while all of the coops are custom, one-off designs, each with a distinct look and function, many of the materials and techniques used are the same.

This chapter is intended as a reference covering the primary supplies and standard building techniques used throughout the book. Many of the projects will refer you back to this chapter for detailed discussions on specific processes, such as installing roofing, doors, or mesh.

DIFFICULTY LEVEL

Each coop in the book is rated for difficulty, to give you an idea of the skill level required for the construction steps. Several factors were considered for each rating, including the complexity of the design, the level of detail and precision needed during fabrication, the size of the overall project, and the sourcing of materials. We don't discourage a beginner carpenter from taking on an advanced coop, but do keep in mind that doing so may drag out the process and require additional research and troubleshooting. Knowing your learning style and the types of challenges you like to take on is a good baseline for choosing what type of coop to build.

YOUR TOOL KIT

Listed on this and the following page are the basic hand and power tools you should own or have access to for building most of the coops. If there's something in the list that you don't have, read through the project you'd like to build to see if the tool is needed. For example, not every project calls for a rubber mallet or a jigsaw.

The list of optional tools includes several woodworking standards — power tools that are ideal for certain operations or that might allow you to add some custom detailing on an otherwise simple feature. In some cases, it might be worth it to borrow or rent one of these or visit a friend who has one in their shop.

If you're hoping that building one of these coops will serve as a gateway to future carpentry and woodworking projects, we strongly recommend investing in high-quality tools. The adage "you get what you pay for" certainly applies to hand and power tools, and having quality tools can save headaches and wasted time throughout the building process.

HAND TOOL ESSENTIALS

- tape measure
- pencils and permanent felt-tip markers
- square
- level
- hammer
- wood chisels
- utility knife
- rubber mallet
- tin snips or aviation snips
- wire cutters
- handsaw (we prefer the Japanese pull saw)
- sliding T-bevel
- chalk line
- clamps
- sawhorses
- painting/wood finishing supplies

Japanese pull saw · chalk line · wood chisels · clamp · square · tin snips · sliding T-bevel

POWER TOOL ESSENTIALS

- circular saw
- miter saw or compound miter saw
- jigsaw
- drill/impact driver with drill bits and screwdriver tips

OPTIONAL POWER TOOLS

- table saw
- router
- orbital sander
- angle grinder
- reciprocating saw
- 12" planer

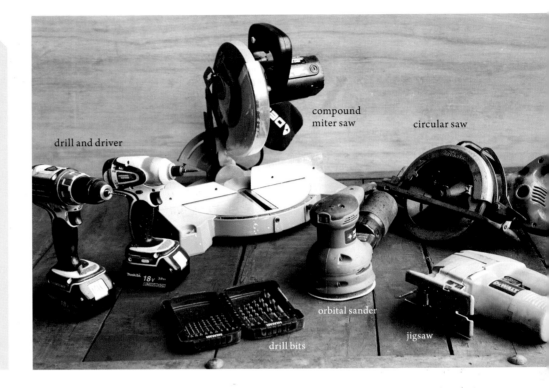

drill and driver

compound miter saw

circular saw

orbital sander

drill bits

jigsaw

A 12-inch planer

Power Tool Safety

We were always taught that one should have a healthy fear when using power tools. They are scary instruments, capable of serious harm. Accidents often happen when you are tired, working too quickly, or operating well above your skill level. It's also important to have the right *approach* with power tools: let the saws do the work, operate them carefully, and don't force them into your material. Always get saws going at full speed before starting a cut. As with kitchen knives, sharp tool blades and bits are safer than dull ones.

When done safely and properly, operating power tools can be an empowering skill set that one has for life. Of course, ear and eye protection are essential at all times, as is a dust mask when there are fine particles flying around. Wear a proper cartridge respirator when working with spray equipment and products with harmful vapors.

From left to right: Plywood, 2×2, 2×4, redwood 2×4

COOP MATERIALS

Generally speaking, chicken coops can be built with ordinary materials available at your local home center or lumberyard. Coops are also a great opportunity to use salvaged, recycled, or repurposed materials, which can lend some personality and uniqueness to a structure, in addition to making good use of existing material. Whatever building supplies you use, remember that a coop must be built for all-season outdoor exposure. That means every screw, nail, hinge, latch, and other metal hardware must be corrosion-resistant. We like to use hardware labeled "galvanized," which has a thicker and longer-lasting finish than hardware labeled "zinc-plated." Wood should be protected with paint or another exterior finish unless it's a naturally rot-resistant species, such as redwood or cedar.

Lumber and Plywood

We use standard, or "common," lumber for most of our coop structures. In North America, the wood species used most often for standard lumber are pine (of various types) and Douglas fir; whatever's most plentiful in your region is what you'll most likely find through local suppliers. Standard lumber is cheap, readily available, and suitably strong for coop framing. However, wherever there is contact with the ground, standard lumber is prone to rot, so here we always suggest using redwood or cedar, which have natural resistance to rot and insects. Conventional builders typically use pressure-treated lumber for these applications, but this is not ideal because the chemicals used in manufacturing pressure-treated woods are unhealthy for the craftsperson as well as for the chickens.

For the best decay resistance, it's important to use heartwood, or "all-heart," grades of redwood or cedar. This lumber comes from the dense core of the tree and

The Container Coop was sided with redwood planks salvaged from an old fence.

is much more rot-resistant than some lower grades, which may contain only some or no heartwood. All-heart lumber is more expensive than other grades, so you may want to use it only for the most critical areas.

When it comes to a flat surface with sheer strength, nothing matches plywood. Usually sold in 4 × 8-foot sheets in thicknesses from ¼" to ¾", plywood is ideal for creating flat surfaces, reinforcing framing, and many other applications. Plywood's structural integrity is due to a series of thin layers of veneer laminated perpendicularly on top of one another, which alternates the grain and results in a very strong and stable board. With most types of plywood, the layers are bound together with glue containing formaldehyde, a known carcinogen. Be sure to take proper precautions to avoid breathing sawdust, or look for plywood products made without formaldehyde.

For plywood that will be exposed to moisture, use marine plywood, which has extra protection against the elements. Some of the coops have details that also call for Euro-ply, a high-quality plywood used for cabinetry and furniture. Euro-ply is more expensive than most standard plywood, but it offers a more uniform end grain without voids, as well as a flatter and smoother surface.

Sourcing and Integrity of Materials

One of the advantages of building a chicken coop is that, because of its small size, you typically don't have to comply with any building codes. This leaves you free to build it how you wish and with whatever materials you wish. This freedom offers the opportunity to source materials through alternative channels, which can provide a more cost-effective, less wasteful coop. Salvage yards are an excellent resource for used common lumber that is typically much cheaper than buying new material. Additionally, there are many Internet resources for finding materials that are otherwise going to the landfill.

We also try to consider the overall life cycle of the product we are using, and what it might look like in, say, 50 years. We do make plenty of compromises (plastic corrugated roofing vs. galvanized metal roofing, for instance), but when given the choice, we try to make a responsible decision without sacrificing the overall design. Sometimes when using found or salvaged material, the finish quality does not have that sparkling new feel, but we encourage folks to embrace the story that salvaged materials can tell, and not hide their imperfections. To further reduce costs of a construction project, try buying things you are going to use a lot of (such as screws) in bulk. Better yet, pool resources with your friends and neighbors to buy and transport materials such as lumber, hardware cloth, and fasteners in bulk.

Roofing

Our choice of roofing for many of our coops is corrugated sheet roofing in either galvanized metal or plastic, both of which come in 2-foot widths and in lengths from 8 to 16 feet. When installing roofing, always slip the lower pieces underneath the layers above so there's no seam for water to penetrate.

Depending on the height and pitch of the roof, you can either pre-cut the pieces or cut them in place. Tin snips are adequate for cutting metal roofing, but a circular saw and a metal cut-off blade produces a cleaner cut edge. For plastic roofing, snips are your best bet. When laying out roofing side by side, be sure to overlap the pieces so the corrugations match nicely. Sometimes different brands of roofing have different corrugations, so it's best to get them all from the same source.

Fasten corrugated roofing with specialty roofing screws; these are equipped with a neoprene washer that seals the screw as it is tightened to prevent water from getting through the hole. These screws must be installed carefully, so the neoprene washer creates a tight seal, but not so tightly that the washer is crushed. Roofing screws are often available with a self-tapping tip, so no predrilling is necessary. To eliminate gaps where the roofing meets the coop walls (where critters could enter your coop), we add "wiggle board," which matches the corrugations of the roofing material and seals the small openings.

For some of the coops, we simply used one piece of flat sheet metal. This is nice because there is no seam to seal and it offers a clean-looking cap. Most metal retailers will cut a sheet of metal to size for you. Galvanizing, painting, or powder-coating your sheet metal will provide many years of rust-free service. Or, if you're like us and don't mind the look of rust, leave it uncoated and let nature have its way.

Hardware Cloth (Mesh)

As discussed in Chicken Coop Essentials, security and ventilation are essential features of a well-designed chicken coop. Most coops include some sort of metal mesh to serve these purposes. We favor a tough type of galvanized metal mesh called hardware cloth, which is a grid of bonded wires with ½" square openings. You can find it in the garden or landscaping section of any home center, where it's commonly sold in 36"-wide rolls of 25-, 50-, and 100-foot lengths. We use 18-gauge mesh on most of our coops.

We can't stress enough that hardware cloth is superior to conventional chicken wire. The openings in chicken wire are large enough to permit the entry of small

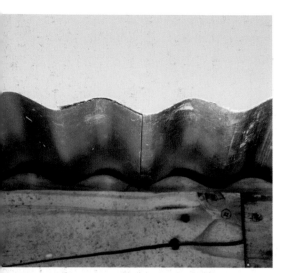

Corrugated roofing overlapped at a seam and mounted on wiggle board at the edge of the coop frame

Hardware cloth sandwiched between 2×4 framing and 1×2 strips

rodents, such as mice, and the wire is thin enough to be torn by some predators. Therefore, we don't recommend using chicken wire for any coop enclosure.

Cut hardware cloth to size by marking it with a straightedge and a felt-tip marker, then cutting with tin snips or aviation snips. Mesh taken straight from the roll can be unwieldy because it tends to curl up, so it helps to flatten it first by bending it backward. Always wear gloves, long sleeves, and safety glasses to protect your body from the sharp points of the cut wire.

For a solid attachment that prevents raccoons and other animals from prying open the mesh, we like to secure the hardware cloth over the coop framing with strips of 1×2 framing lumber and screws. If you put several screws through the mesh openings into the frame of the coop, an ultra-tight seal is made that will keep your chickens safe. This technique is more work than simply stapling the mesh (which is okay, provided you use enough staples), but in the long run it makes for a much more sound coop and peace of mind for the owners. In some cases (such as the Container Coop, on page 182) we use a heavier gauge of mesh that comes in a sheet form and attach it with heavy-duty U nails, which also makes for a tight seal.

Locks and Latches

An indispensable finish detail for any coop door is making sure it locks tightly so that predators can't get in, especially at night. Raccoons are particularly industrious carnivores that can exploit design flaws that leave the coop vulnerable. Because of this, there are a few latches we favor: the spring-loaded hook-and-eye latch, the gravity latch, and the swivel gate latch. All of these have a hole that you can put either a lock or carabiner through to secure it. Barrel-bolt latches are cheap and easy to install, but use a heavier-duty latch to lock the hens in at night.

ESSENTIAL TECHNIQUES

This section covers many of the common building techniques used time and again throughout the book. Some are standard woodworking procedures that can apply to any kind of project; these will be most helpful to beginning builders. Other techniques are specific to coop construction and will be helpful for anyone creating one of the coop designs in the book.

Spring-loaded hook-and-eye latch

Gravity latch

Swivel gate latch

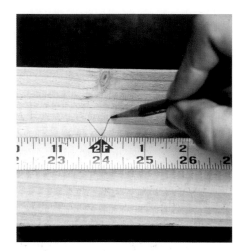

Marking the cut location with a caret

Drawing the cutting line with a square

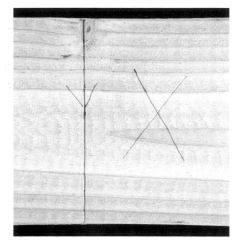

A finished cutting line with an X denoting the waste side of the cut

Measuring and Marking

Laying out a cut properly before you actually cut it will save a lot of frustration down the road. Because wood from the lumberyard is not always square on the ends, it's a good idea to cut off a little bit of one end on a miter saw to ensure that it is square; this is the end you will measure from to make the final cut.

1 Using a pencil and tape measure, mark the cutting line location by making a V (called a *caret*), with the point of the V at the exact measurement.

2 Set the point of the pencil on the Vs point and slide a square right up to your pencil, then mark the cutting line across the face of the board. The square ensures that the line is straight and is perpendicular to the board's long edges.

3 Remember to line up the cut so that the thickness of the blade is on the waste side of the cutting line. This ensures that the *kerf* — the thickness of the blade or, rather, the material removed by the blade — is taken off the waste piece and not the workpiece.

Fastening

When it comes to fasteners, we always prefer using screws to nails. Screws create a stronger joint, they're usually easier to install, and they allow for easier disassembly, in case you have to break down your building to move it. Whenever putting in a

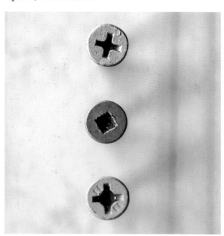

Screw head types from top to bottom: Phillips, square, combination

screw, particularly near the end of a board, it's important to drill a pilot hole before driving in the screw. This will prevent splitting the wood and make it easier to drive the screw. A 3/32" drill bit is a good size for pilot holes for deck screws. As a rule of thumb, the piloting bit should be slightly smaller in diameter than the screw's shank — the solid center of the screw, around which the threads run. This allows for the screw to grab the wood tightly. A nice drill bit to use is the combination pilot-countersink bit, which allows your screws to be neatly installed flush with the wood surface.

We always use deck screws, which are made for outdoor projects and have a coating that prevents rust. Instead of standard Phillips-head screws, we strongly recommend using square-drive or combination-drive screws and a square-drive driver bit on your drill or impact driver. Square drive offers better resistance to stripping, making it far superior to Phillips and preventing serious stress and headaches. While not necessary, having an impact driver in addition to a drill is handy. Having both means you can have the drill set up for predrilling, and use the impact driver to drive in screws. Another feature of the impact driver is that its hammering action does the work, rather than your wrist, and allows for easier penetration by the screw.

For some of the coops, we specify *toe-screwing* — fastening two perpendicular pieces by driving the screws at a 45-degree angle. Toe-screwing is useful for situations where you can't screw straight into your material. Always drill pilot holes when toe-screwing.

Driver types: Phillips (top) and square (bottom)

Toe-screwing

STRAIGHTEDGE GUIDES FOR HANDHELD SAWS

To make clean, straight cuts with a circular saw or jigsaw, use a straightedge guide. Start by measuring the distance between the outside of the saw blade and the side edge of the base, or "foot," of the saw (be sure to measure to the outermost point of the saw tooth, not the smooth part of the blade). Mark a line on your material parallel to your cutting line but offset by the distance measured on your saw. Clamp a straight board (a piece of hardwood or the factory edge of a plywood or MDF sheet works well) or a level to the workpiece in at least two places so its edge is on the offset marked line. Make the cut with the saw's foot riding along the straightedge.

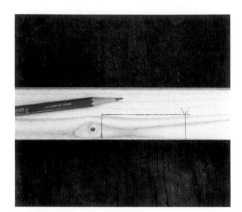
Marking the outline of a hinge

Chiseling out the wood

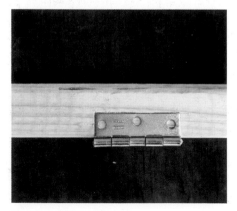
Hinge ready for screws

Doors and Hinges

Hanging doors and installing hinges can be tricky, but with the proper setup it can be done smoothly and without headache. We usually build our door frames on the inside and in the same plane as the rest of the coop framing, with a ⅛" to ¼" gap on all sides. Usually it's a good idea to add triangular bracing to your door frame to prevent sagging on the side without hinges.

When installing hinges, it helps to make a mortise so that the hinge leaves are set into recesses in both the door and the frame. This will minimize the gap and allow for a cleaner, tighter fit inside the door opening. The traditional way to make a mortise is by tracing the outline of the hinge leaf on the side of the door and the door frame. Using a sharp chisel and a mallet, score the perimeter of your mortise with the tip of the chisel, and then pare away the wood until it is deep enough to accept the hinge. It is usually only about ⅛" deep, so this shouldn't take too long.

A more advanced mortising method is to use a handheld router. You can make a jig to guide the outside of the router bearing (using a bearing-guided straight bit) and set the depth of the bit to cut away the precise thickness of the hinge leaf. To avoid climb-cutting, always start the router away from the material and move in the correct direction.

Finding Angles

Many of our coops have irregular angles in them, requiring some fancy cutting. Don't worry if your geometry skills are rusty; there are tools to help with the cuts. For angles from 45 to 90 degrees, the miter saw is the idea tool to use, but it can cut only up to a certain size of material. Otherwise, it's usually best to make angled cuts with a circular saw, following a fully marked line.

Rather than tackling tricky equations, a sliding T-bevel will allow you to record the angle you want to cut so you can set your miter saw or mark your cutting line accordingly.

1   Set the T-bevel at the desired angle by holding the handle against the edge of the board and adjusting the blade (the metal tongue) to follow the angle, then tightening down the blade with the wing nut.

2   Take the T-bevel to the miter saw. Put the handle against the fence and pivot the saw so the blade rests against the bevel. Be careful not to activate the saw when you are adjusting the angle.

3 Test-cut your angle on a piece of scrap wood to make sure it is right before cutting your final piece.

If you have access to a table saw, you can tilt the blade to the desired angle, or use a miter gauge and set the piece at an angle in relation to the vertical blade. For compound angles — cuts that are mitered *and* beveled, or angled on two planes — you'll need to use a compound miter saw, which lets you adjust the tilt on the blade in two axes. These are challenging cuts to make, and we recommend doing test cuts to make sure your saw is where you want it. Most new miter saws have compound cutting capability, and these are commonly available for rent at home centers and rental outlets (renting a saw for a day is a very good idea if you have a lot of angled cuts to make). The Chick-in-a-Box (page 94) and Cupe (page 168) projects involve compound miters.

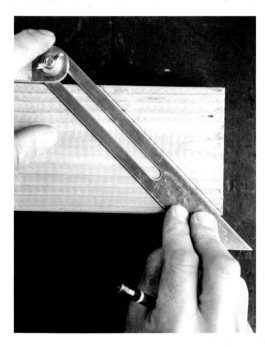

Matching a marked angle with a sliding T-bevel

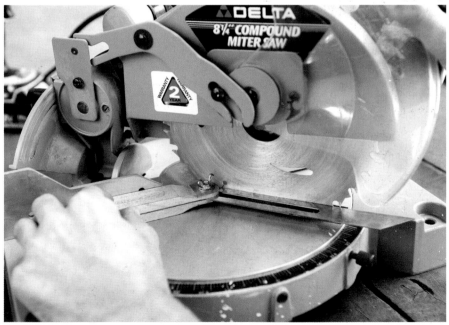

Adjusting the miter saw blade to match the T-bevel's angle

Cutting Notches

Many of our coops call for notches to create a nice, tight-fitting joint. When preparing to make a notch, first mark the depth and width of the notch using a pencil and a square. There are several ways to make a good notch. If using a circular saw:

1 Start by setting the blade to cut the full depth of the notch.

2 Make two cuts, one on each side of the notch to form the outside edges.

3 Make a series of cuts to remove most of the material in between.

4 Finally, use a chisel to remove the remaining wood and smooth the bottom of the notch.

A similar procedure can be done with some higher-end miter saws that allow you to set the depth of cut. You can also do the same thing with a nice handsaw, by using your eye to cut down to the line.

If you have access to a table saw, notching can be done much easier and faster. You can either use a regular blade or a *dado blade,* which essentially is a set of blades stacked together to cut a wide kerf, removing up to ¾" in one pass.

1 Install a dado blade as directed by the blade and table saw manufacturer.

2 If you are making several crosscuts that are exactly the same size, you can set a jig by clamping a scrap board to the rip fence just in front of the blade, to establish a clearance gap between the workpiece and the fence while it is going through the blade.

3 Push the workpiece through the cut with a miter gauge or sled.

4 For wide notches, make two initial outer cuts, as with the circular saw, then cut away the interior waste material to complete the notch.

Making Joints: Miters, Dadoes, and Half-Laps

When it comes to joining two pieces of wood that are perpendicular to each other, there are many possibilities, some of which are for structural purposes and some more for aesthetics. Sometimes, on an exposed joint where we don't want the end grain visible, we'll use a miter joint. A *miter joint* is two pieces of wood cut at 45-degree angles that join to make a 90-degree corner. Miter joints on trim and frames are a nice touch and give a professional look. Make several test cuts to be

Chiseling out the remaining wood

With the saw off, lining up a dado cut using a sled (*left*) and pushing it through (*right*)

sure your saw is as close to 45 degrees as possible; otherwise there will be gaps in your frame, particularly at the corner that is assembled last. Miter joints are easiest to cut on a miter saw, but you can also use a handsaw (typically a backsaw) and an inexpensive plastic or metal miter box.

A *dado joint* is another classic woodworking joint that we employ in some of our coops, usually for nesting boxes and the like. A dado is a groove cut in a piece of wood that encloses the end grain of the adjoining piece. A dado that runs along the edge of a piece is called a *rabbet*. Both dadoes and rabbets can be cut on the table saw with a stacked dado blade (see Cutting Notches, facing page) or with a router and a straight bit.

An assembled miter joint

An assembled dado joint

An assembled rabbet joint

Cutting a rabbet with a router and straight bit along a straightedge clamped to the workpiece

Half-lap notches

An assembled half-lap joint

A half-lap corner joint

Another strong and decorative traditional joint we like to use is the *half-lap*. This is comprised of two identical notches cut into the mating pieces. The notch depth is half the thickness of the material so that the pieces are flush when the joint is assembled. You can cut the notches for a half-lap joint using the same techniques described in Cutting Notches (page 32). Half-laps can be made anywhere along the mating pieces or to create corner joints.

Always make some test cuts on scrap wood to ensure the notches are the right depth and width. Since both pieces get the same notch, any errors in the sizing are doubled.

Plunge-Cutting

Plunge-cutting is a handy technique with a circular saw that enables you to begin a cut on the interior of a workpiece, such as when making a window cutout on a sheet of plywood. This is a safe operation when done properly but is extremely dangerous if done improperly.

Here's the right way to make a plunge cut:

1 Mark your cut clearly.

2 With the saw off and the blade completely stopped, tilt the saw forward and position the front edge of its foot (base) squarely on the workpiece.

Positioning the saw for a plunge cut

3 Retract the blade safety guard and align the blade over the cutting line, making sure the saw is far enough from the end of the line so that it won't overcut when the saw is horizontal and fully engaged in the wood.

4 Keeping the blade guard retracted, start the saw and let it come to full speed. Slowly tilt the saw down to its normal horizontal position, engaging the blade into the wood as you level the saw.

5 Once the foot is flat against the workpiece (fully horizontal), move the saw forward to complete the cut, as with the normal operation. Keep the saw on throughout the operation. Let the safety guard move when it will no longer interfere with the cut.

6 To prevent overcutting at the corners of a cutout, stop the circular saw cut just before the corner. Because the blade is round, it can't reach the full depth of the material, so you have to finish the cuts at the corners with a handsaw or a jigsaw.

Plunge-cutting can be dangerous: Due to the blade's rotational direction, a circular saw exerts a backward force during a cut. If you plunge-cut into the material too quickly or without a firm grip on the saw, the saw can potentially jump backward. Since you have the blade guard retracted, there's additional potential for the blade to cause damage — to the workpiece or to you. Again, this is an easy and safe operation once you get the feel of it. Just be sure to practice a few times on relatively thin material, such as plywood. You can plunge-cut into thicker materials, but the backward force of the saw is relatively stronger, due to more of the blade engaging the wood.

The alternative to plunge-cutting when making an interior cutout is to use a jigsaw with a starter hole for the blade. Drill a ⅜" hole inside the waste area, then insert the jigsaw blade into the hole to begin the cut.

Initiating the plunge cut

Completing the plunge cut

Finishing the corners with a handsaw

Bracing a post plumb, ready for concrete in the tube form

"Placing" Concrete

A wise builder we once worked with told us, "You don't *pour* concrete; you *place* it." So any time concrete is mentioned, we always note the distinction. In this book, concrete is used exclusively for installing structural posts. Here's what you do:

1 Start by digging a hole at least 2 feet below the frost line, so the concrete will rest on undisturbed soil. If the hole is not deep enough, your post can heave from the ground when the ground freezes. Typically, 3 to 4 feet below ground level will suffice. The hole's diameter should be at least twice the width of the post, plus 2" or 3". You can use a post hole digger and a spade shovel to excavate material. **NOTE:** *Call before you dig. Make sure there are no underground utility lines in or near the coop site before you break ground.*

2 Add 4" to 6" of dry gravel to the bottom of the hole and tamp it down.

3 You can use the hole itself as the form for the concrete, but we like to use cardboard concrete tube forms (such as Sonotubes), which you can buy at home centers and lumberyards and cut to length. The diameter of the form should be at least twice that of the post. Set the form into the hole and fill around the form with soil, checking the form with a level to make sure it's plumb and tamping the soil firmly as you go.

4 Set the post into the hole and brace it so it's perfectly plumb, using two 1×2 cross braces.

5 To mix the concrete, pour the contents of one bag into a wheelbarrow or trough and make a hollow in the middle to receive the water. Start adding water a little bit at a time, taking breaks to mix it thoroughly. It helps to have two people doing this, one mixing from each side with a hoe or shovel. You'll notice that even though it looks wet on top, at the bottom there still will be plenty of dry concrete. Keep mixing and adding water in small increments until the mixture is consistently wet. Most importantly, don't add too much water and make it soupy, which weakens the concrete and adds to the curing time.

6 When the concrete is fully mixed, place it in the form or hole a little at a time, taking breaks to tamp it down with a scrap piece of wood. Alternate sides to fill the hole evenly all around. Overfill the hole around the post, then form a little dome at the top to direct water away from the post.

USE AN ANGLE GAUGE

You can use a sliding T-bevel or a piece of scrap wood cut to the desired angle to serve as a gauge for testing the bent angle. This way, you don't have to remove the metal from the brake each time to test the angle.

Bending sheet metal with a homemade brake

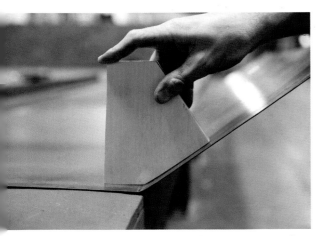

Checking the angle with a custom-cut wood block

7 Confirm that the post is plumb, and let the concrete cure for at least one day before removing the braces. Different types of concrete and applications have different curing times, so be sure to follow the manufacturer's directions.

Bending Sheet Metal

Some of the coop designs call for a roof made with a single piece of sheet metal, with a crease bent into it. Others require bent pieces of metal for flashing or other roof elements. A metal fabricator would use a hydraulic brake to make such a crease, but with thin metal you can do the same with a homemade brake.

1 Place the metal sheet on a sturdy work table so the line of the bend is at the edge of the table.

2 Clamp a straight 2×4, 4×4, or piece of steel tubing on top of the sheet, aligned with the bend line.

3 Bend the sheet metal against the edge of the table and/or the board. If necessary, clamp an additional straight edge to the end of the sheet for better grip, as shown in the photo at left.

Sanding and Finishing

Usually when you're done building a coop, there are pencil marks, dirt, and chalk line residue on the exterior. It's always nice to give it a sanding to clean it up and make everything smoother to the touch. We prefer using an orbital sander, starting with a coarse grit (60 or 80) and working up to a finer grit (150 or 180), so be sure to wear a respirator. After sanding, thoroughly wipe off the dust and apply the finish of your choice.

There are many different finishing products out there, ranging from natural oils to polymer-based formulas. Because the chickens are producing food, products with lower toxicity are preferred. A good finishing product will not only seal the wood to protect against water and UV light damage and make it easier to clean, it also darkens or colors the wood to make it "pop" visually. Keep in mind that a lot of clear coats do not offer UV protection, so if the coop will be exposed to a lot of sunlight, something more protective like paint or a pigmented stain is a good option. Follow all product directions, and apply multiple coats for the best protection and appearance.

Making Mistakes

One final thing to keep in mind as you venture into coop building: Mistakes are inevitable, so don't be afraid to make them. We've found in our many project endeavors that solving problems on the fly is one of the most stimulating parts of making things, and seeing mistakes as opportunities rather than personal failures makes the experience much more fulfilling. If you're struggling with a difficult problem, taking a break and coming back to it another day is often all that's needed for a solution to come to mind. This can also be a great opportunity to consult a friend or other skilled person who can help. You'll find that most people like to share their expertise and opinions, and this could lead to a deepening of existing friendships or entirely new ones.

COOP
PROJECTS

TWO SIMPLE CLASSICS

To start off the projects with a couple of easy-to-build and versatile designs, we challenged ourselves to create simple and inexpensive, yet elegant, coops using only basic hand tools and a circular saw. Both of these coops use the same raised 4×4 platform design, and you can decide which "house" structure to add from there. Either can be built in a weekend.

THE STANDARD

THE A-FRAME

MATERIALS

FOR THE A-FRAME COOP (INCLUDING PLATFORM)

> One 4-foot rot-resistant 4×4
> Six 8-foot 2×4s
> Two 4 × 8-foot sheets ½" plywood
> Two 10-foot 2×4s
> Three 8-foot 2×2s
> Sixty 2½" deck screws
> Seventy-five 1¼" deck screws
> Twenty-five 2" roofing screws with neoprene washers
> Four square feet ½" galvanized hardware cloth, 18-gauge
> Four 2½" hinges with screws
> Finish materials (as desired; see step 8)
> One 5 × 10-foot sheet mild steel, 16-gauge
> One 6-foot-long 8" × 8" piece V-ridge cap flashing
> Two door handles with screws
> Two locking door latches with screws

ADDITIONAL MATERIALS FOR OPTIONAL INSULATION:

> One 4 × 8-foot sheet and one 4 × 4-foot sheet ½" plywood
> Twenty-four linear feet fiberglass blanket insulation for 2×4 walls
> Fifty 1¼" wood screws

FOR THE STANDARD COOP (INCLUDING PLATFORM)

> One 4-foot rot-resistant 4×4
> Fifteen 8-foot 2×4s
> One 4 × 8-foot sheet ½" plywood
> One 8-foot 2×2
> Seventy-five square feet siding
> One hundred 2½" deck screws
> Thirty 1¼" deck screws
> Twenty 3½" deck screws
> One hundred finish nails (or trim head or security screws)
> Ten 2" roofing screws with neoprene washers
> One operable window, approximately 2 × 2 feet
> Two 2½" hinges with screws
> Two 2½" hinges with screws (as needed; see step 4)
> Twelve 2" round louver vents
> Twenty-five square feet roofing material (see options in step 9)
> Two 2" security hasp latches with screws
> Two door handles with screws

ADDITIONAL MATERIALS FOR OPTIONAL INSULATION:

> Three 4 × 8-foot sheets ½" plywood
> Fifty linear feet fiberglass blanket insulation for 2×4 walls and roof
> One hundred 1¼" wood screws

SPECIALTY TOOLS

> Circular saw with abrasive metal cutoff blade (for sheet metal roofing, as needed)
> 2" hole saw (for Standard Coop)
> Respirator or mask and gloves (for cutting and installing insulation)

Weatherizing Tips

Both house structures feature 2×4 stud walls spaced to accommodate standard home insulation, so they're easily adaptable for hotter or colder climates. We've included supplies and steps required for adding optional insulation.

BUILDING THE PLATFORM

Both the A-Frame and the Standard Coops start with this platform. To make it, we cut up a 4×4 redwood post into 12" sections to use as legs. If you like, you can cut longer legs to raise the coop higher and give the hens a dry space underneath for hanging out.

1. CUT THE PLATFORM PARTS.

From the rot-resistant 4×4, cut four legs to length at 12". To cut a 4×4 with a circular saw, mark cutting lines across all four faces of the workpiece, and set the saw to the maximum depth of cut. Make one cut across one face of the piece, then flip it over and make a second cut from the opposite face. Alternatively, you can use a miter saw with a cutting depth of at least 3½".

Cut two common 2×4s to length at 48"; these are the end joists. Cut four 2×4s at 45"; these are the common joists. Finally, cut the ½" plywood floor deck to size at 48" × 48".

2. ASSEMBLE THE PLATFORM.

Position the two end joists over the ends of two common joists, with all pieces on-edge, to form a 48" square frame. Fasten through the end joists and the commons into the posts with pairs of 2½" screws. You may find this easier to do with the frame upside down on your work table. The tops of the legs should be flush with the top edges of the joists. Install the remaining two common joists in the same way, spacing them at about 16" on center between the outer common joists.

Lay the plywood deck panel over the top of the assembled frame and align the outside edges of the deck and framing; this ensures the frame is square. Fasten the plywood to the interior common joists with 1¼" screws. Don't screw along the perimeter of the deck yet — you will do this when you install the walls.

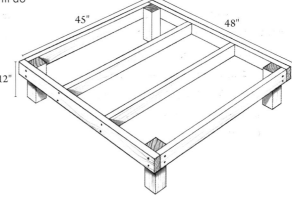

Assembled platform frame

3. INSULATE THE FLOOR (OPTIONAL).

Cut a piece of ½" plywood to size at 48" × 48" to cover the underside of the floor frame. Flip the coop platform upside down. Notch the corners of the plywood to fit around the 4×4 legs. Wearing a respirator and gloves, cut three pieces of fiberglass insulation to length at about 46", using a utility knife and a scrap of wood or a level as a guide. Lay the insulation into the joist cavities; it should fit snugly without being compressed. Cover the framing with the plywood and fasten it with 1¼" screws. Fasten all the way around the perimeter and add a few screws in the middle of the plywood, along the centers of the joists.

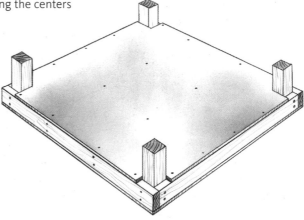

Bottom view of platform with plywood covering insulation

BUILDING THE A-FRAME COOP

A-frames are among the most time-honored and ubiquitous coop designs in existence, and for good reason. They make economical use of materials, they're space-efficient, and they're good for keeping in heat. Initially, we voted against doing an A-frame because there are so many good examples out there, but after building most of the other coops, we felt something was missing. We decided that having a collection of DIY chicken coops without an A-frame was like having your whole body pierced but not your ears.

While keeping the construction as simple as possible, we wanted our coop to model the clean lines of the Nordic A-frame style, which has characteristically steeper angles than other versions. Sheet metal roofing helps to preserve this aesthetic. If it's easier for you to use standard roofing (such as cedar or asphalt shingles or corrugated metal roofing), you can substitute that for the sheet metal. Each side of the coop is the same for ease of construction, and cross-ventilation is provided by meshed "windows" in the gable ends.

1. FRAME THE SIDE WALLS.

The sloping sides of the A-frame house comprise both the roof frame and side walls. They are made with 2×4 rafters that meet at the peak of the structure and 2×4 horizontal supports — called purlins — installed between the rafters. Cut two sets of 2×4 rafters as shown in the *Rafter diagram*, using 10-foot 2×4s. Note that the top end of each rafter is angled at 27 degrees and the bottom end is angled at 63 degrees. Mark these cuts using a sliding T-bevel (see page 30), and make the cuts with a circular saw.

Test-fit the first two rafters on the coop platform, and make any necessary adjustments to the cuts for a good fit at the peak and the platform deck. Then, use the rafters as templates to mark two more matching rafters, and make the cuts. Install the rafters as shown by screwing the pairs together at the peak with two 2½" screws and to the platform with three screws at each joint.

Cut six 2×4 purlins to length at 45". Install the two lower purlins on each side between the rafters, as shown in *Purlin placement*. Position the purlins so they are flush with and perpendicular to the outside edges of the rafters. Fasten the purlins to the rafters with 2½" screws. Install the top purlin on each side so its top face is 7" from the peak of the roof.

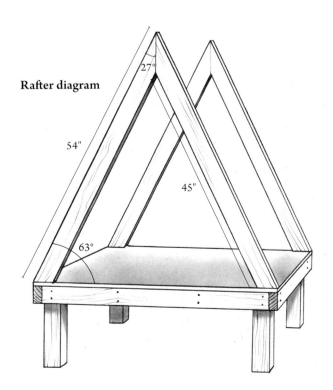

Rafter diagram

27°

54"

45"

63°

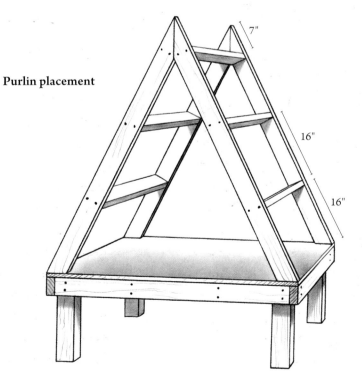

Purlin placement

7"

16"

16"

2. FRAME THE DOOR OPENINGS.

Cut three pieces of 2×2 for each of the two door frames, as shown in the *Door frame diagram*. The top (horizontal) header piece has its ends angled at 63 degrees and installs level with the platform deck. Fasten the door frame pieces together and to the rafters and platform with 2½" screws.

3. INSULATE THE SIDE WALLS (OPTIONAL).

If you'd like to insulate one or both of the sloping side walls, cover the interior wall(s) with ½" plywood fastened to the framing with 1¼" screws, then cut and install insulation to fit snugly into the cavities between the purlins and rafters. The plywood is necessary for cleanliness and to prevent the chickens from picking at the insulation. Make sure the edges of the plywood do not extend beyond the outside faces of the rafters.

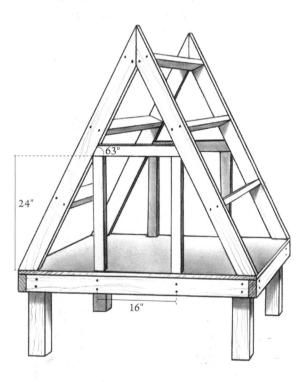

Door frame diagram

4. INSTALL THE SIDING AND VENTILATION WINDOWS.

Cut a piece of ½" plywood to 48" × 32". Position the panel against one of the vertical ends of the coop frame so the bottom and side edges of the panel are flush with the bottom and sides of the platform, and clamp the panel in place. Use a pencil to trace the outline of the coop frame onto the back side of the plywood. Also trace inside the door frame.

Unclamp the plywood and cut along the lines with a circular saw. If you cut out the door opening carefully by plunge-cutting with a circular saw (see page 34) and finish the corners with a handsaw, you can use the cutout piece later for creating the door. Position the siding over the end wall again so all edges are flush and fasten it to the coop framing with 1¼" screws. Repeat the same process to install siding on the other end wall.

Create the frames for the ventilation windows using scrap pieces of plywood. Use the same process of tracing along the coop frame to mark and cut two triangles to cover the end walls between the top of the plywood siding and the roof peak. Make a triangular cutout in the center of each piece that follows the lower edges of the rafters and has a 2¼"-wide strip along the bottom of the triangle, as shown.

Cut triangles of hardware cloth slightly smaller than the outer edge of each plywood window frame. Sandwich the mesh and frames over the rafters and fasten them with 1¼" screws.

Siding and ventilation windows installed

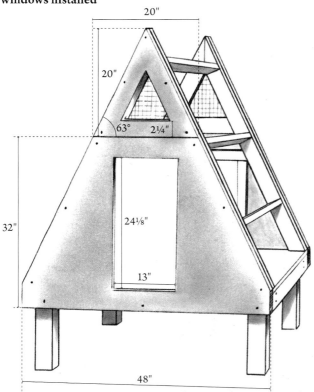

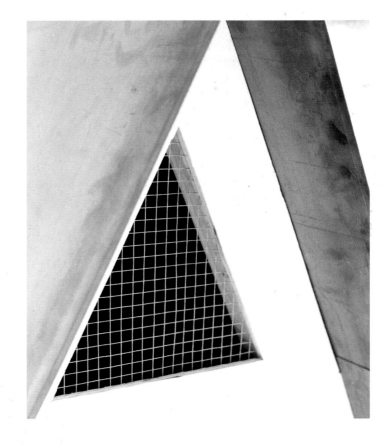

5. BUILD AND HANG THE DOORS.

For each door, cut four pieces of 2×2 to create a frame that's ½" narrower and shorter than the door opening on the coop. Join the pieces with 2½" screws. Use the leftover door piece or cut a new piece of plywood to match the outside dimensions of the frame and attach the plywood with 1¼" screws. Hang each door to the coop with two 2½" hinges so the door is centered within the opening.

6. CREATE THE NESTING BOX.

Cut a piece of ½" plywood for the nesting box panel, as shown in the *Nesting box cutting diagram*, using a circular saw or jigsaw. Also cut four small blocks from the plywood, two at 1½" × 3½" and two at 1½" × 5". Install the panel and blocks, using 1¼" screws, as shown.

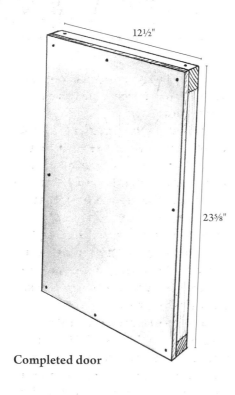

Completed door

Nesting box cutting diagram

3½"
27°
12"
15¼"
10"

Nesting box installed

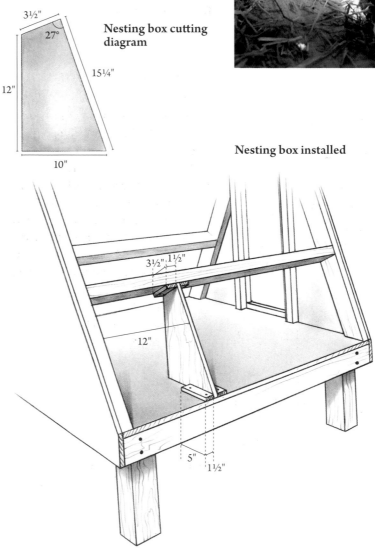

3½" 1½"
12"
5"
1½"

7. ADD THE ROOSTING BAR.

On the inside of the coop, opposite the nesting box, cut and notch a 2×2 roosting bar cleat to fit between the rafter and the door frame, as shown (see page 32 for help with cutting notches). Install the cleat to the framing with 2½" screws. Cut and install a matching cleat on the opposite end wall, then cut a 2×2 roosting bar so it fits nicely into the notches of the cleats.

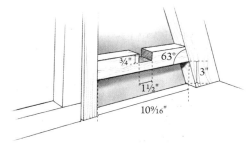

Roosting bar cleat

8. INSTALL THE ROOFING.

If desired, paint or finish the exterior of the coop with the finish of your choice before installing the roofing.

We cut our sheet metal roof into shingles, but it's also possible to install the roofing as two whole pieces with a ridge cap. For the shingle method, start with a 5 × 10-foot piece of steel and cut it in half with a circular saw and an abrasive metal-cutoff blade. Out of each of the two halves, cut the following pieces:

- One at 60" × 23"
- One at 60" × 17½"
- One at 60" × 19½"

Coop roofing installed

Each half of the original full sheet will yield the three different dimensioned pieces with no waste.

To install the shingles, start at the bottom of the coop with the 23" piece. Line it up with the top of the first purlin from the bottom and clamp it in place. Place the 17½" piece so that its top edge is aligned with the top of the second purlin, then drill pilot holes and fasten through both sheets and into the rafters and first purlin with three 2" roofing screws. Position the third shingle and screw along the second purlin and the rafters.

Repeat the process to install the shingles on the other sloped side of the coop. Cut a piece of ridge cap flashing to length at 60" and fit it over the shingles at the roof peak. Fasten the cap with roofing screws driven into the top purlin and the rafters.

Add paint.

Install handles and latches to the doors to complete the coop.

BUILDING THE STANDARD COOP

The Standard Coop is like a modified cube in form and carries some simple custom details that give it the look of a modern classic. The simple shed-style roof slopes to one side and overhangs all of the walls to create both decorative and practical eaves. There's a door at the back, while the opposite wall features a hinged window to brighten the interior and give the birds a room with a view. This coop should inspire you to use your imagination when it comes to materials, or perhaps mix and match elements from other coops in this book. The window for our coop came from a friend's home remodel, the redwood siding was from an old fence that someone took down, and we used a sheet of 14-gauge steel to cover the roof (but it would easily accept corrugated metal, another type of sheet metal, or maybe a big street sign).

1. FRAME THE WALLS.

The Standard coop has two side walls that are framed with 2×4s, much like ordinary house walls. One of the walls is 6" taller than the other, and the difference in heights creates the roof slope.

For the high side wall, cut two 2×4s to length at 48"; these are the horizontal plates at the top and bottom of the wall frame. Cut four 2×4 studs to length at 30". For the low side wall, cut two plates at 48" and four studs at 24".

Assemble each wall by placing the plates over the ends of two studs to form a rectangular frame. Screw through the plates and into each stud with two 2½" screws at each joint. Install the two remaining studs in each wall in the same way, spacing the studs about 16" on center.

Position each wall along a side edge of the completed platform and fasten the bottom plate to the platform with 2½" screws.

Side walls installed

2. CUT AND INSTALL THE RAFTERS.

Rough-cut four 2×4 rafters to length at 54". To mark the "bird's mouth" cuts that allow the rafters to sit atop the side-wall plates, as shown in *Position of rafter for tracing*, position one rafter against the ends of the side walls so that its bottom edge is ¼" above the inside, bottom edge of each wall's top plate. The ¼" offset allows the interior plywood wall covering to overlap onto the plates, eliminating any gaps.

Use a pencil to trace along the top and both side edges of each wall plate, marking their outlines onto the back face of the rafter. Remove the rafter and cut along the lines with a circular saw or jigsaw. Test-fit the rafter on the walls and make any adjustments necessary for a good fit. Then, use the cut rafter as a pattern to mark the remaining three rafters, and make the cuts.

Install the rafters onto the wall plates using 3½" screws. Position the two outside rafters flush with the ends of the walls, and space the two interior rafters about 16" on center in between.

3. SHEATH AND INSULATE THE COOP INTERIOR (OPTIONAL).

If you're insulating the walls and roof, interior sheathing is required. Even without insulation, sheathing provides a nice, flat surface for painting and easy cleaning.

Cut ½" plywood to fit for each side wall. The panels should extend from the coop floor to the bottom edges of the rafters and be flush with the wall studs at both ends. Install the panels with 1¼" screws, as shown in *Rafters installed*.

Cut a third plywood panel for the ceiling so it abuts the wall panels and is flush with the outside faces of the outer rafters. Install the panel with 1¼" screws; it helps to have a buddy or some clamps for this. If desired, fill the wall-stud and rafter cavities with insulation.

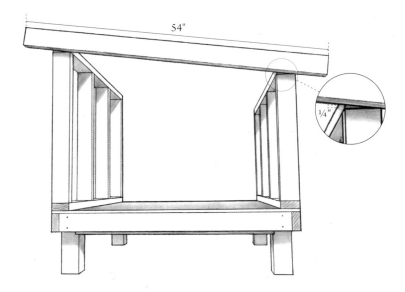

Position of rafter for tracing

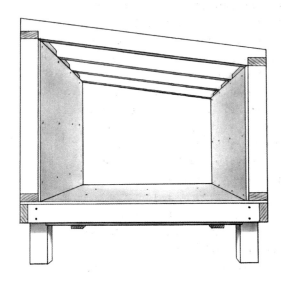

Rafters installed (with interior sheathing)

4. INSTALL THE WINDOW.

Build a frame for your window, using 2×4s: Cut a horizontal header for the top of the frame, spanning between the side walls, and install the header with 2½" screws. Add a vertical stud at each side of the frame so the opening is centered on the wall and is about ½" wider than the window. Install the window in the frame as appropriate for its design. If the window itself isn't operable (just a fixed sash, for example), hang it to one of the side studs with two hinges; it's beneficial that the window can be opened to provide cross-ventilation inside the coop (see pages 18 and 30 for tips on cleaning access and hanging windows).

5. FRAME AND FIT THE DOOR.

The door is custom-built and can be any size you like. Our door opening measures 13½" × 24" and leaves about 12" of solid wall at each side for adding insulation. As with the window, frame the door opening with a 2×4 header and two 2×4 studs, centered on the wall horizontally. It's best not to have a threshold at the door opening, so the coop is easy to sweep out.

Build a a simple rectangular door frame with four 2×4s and 2½" screws. The outside dimensions of the frame should be ½" narrower and shorter than the door opening in the coop wall.

Position the door frame into the opening, using shims to set an even ¼" gap at all sides. Mark the locations for two 2½" hinges onto the door frame and one stud of the opening. Mortise the door frame and stud so the hinges will mount flush (see page 30). You will hang the door along with the siding installation (step 7).

6. INSULATE THE END WALLS (OPTIONAL).

Cut ½" plywood to cover the interiors of the end walls, making cutouts for the door and window. Install the sheathing with 1¼" screws. Wearing a respirator and gloves, fill the framing cavities with insulation.

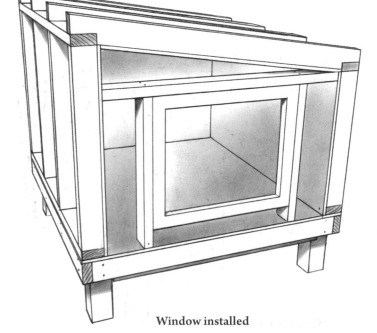

Window installed

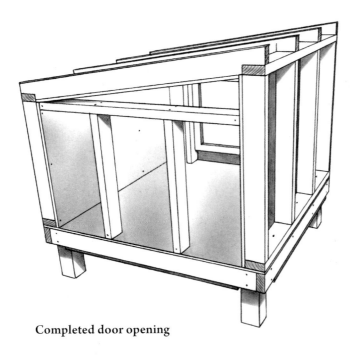

Completed door opening

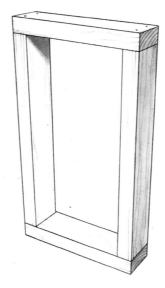

Door frame

7. ADD THE SIDING.

Install siding on all four walls of the coop. We used finish nails to attach the siding, since they have small heads. Trim-head or "security" screws are also good options.

At the top of the high side wall, run the siding up so it's flush with the tops of the rafters to eliminate any gaps when you add the roof. At the low side wall, stop the siding about ¼" below the tops of the rafters to allow clearance for the sloping roof. At the wall corners, run the siding a little bit short of the side edges if you are covering the corners with trim. If you're not using trim, you can run the siding long beyond the edges and trim it neatly with a circular saw after the siding for each wall is installed.

For corner trim, we cut pieces of leftover siding and screwed it in place over the wall corners. Depending on the width of your siding material, you may want to rip the pieces to the desired width. Alternatively, you can use 1×2, 1×3, or 1×4 lumber for the trim. Cut the top ends of the wall trim to follow the roof slope. Also add trim around the window and door openings, if desired. Cover the door frame with siding and hang the door.

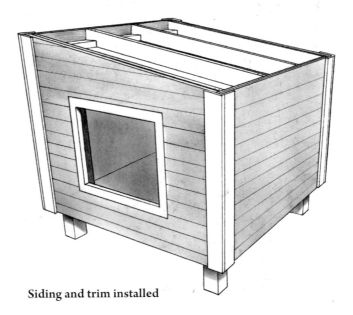

Siding and trim installed

8. INSTALL THE VENTS.

To provide some ventilation when the coop is closed up, we added 2" louver vents, which you can find at any large hardware store or home center. They require a 2"-diameter hole cut with a hole saw into the siding from the outside and through the plywood on the inside (if you sheathed the interior). We used three vents per wall on each side of the coop (not on the end walls), placing them 6" below the tops of the walls and centering them between the wall studs. If you sheathed the interior, you'll need 12 vents total.

It's a good idea to make a test cut in scrap plywood, using the hole saw, and test the fit of the louver — it should pop in with just minor pressure. Once you know you have a good fit, mark and cut your holes in the coop walls, cutting slowly to prevent tearing out the siding, then pop in the vents. If your walls are insulated, the hole saw should remove the insulation between the two vents. Alternatively, you could cut it out with a knife after the holes are drilled in the siding.

Louver vents installed

Roofing installed

9. ROOF THE COOP.

For our roof, we bought a 5 × 10-foot sheet of 14-gauge mild steel from our local supplier and cut it down to a 5-foot square with a circular saw and an abrasive metal cutoff blade. Your local sheet metal shop can cut a piece for you, if you prefer. Other suitable metals for the roofing include aluminum, stainless steel, or copper, but all of these cost more than standard steel, which will rust if not given a protective coating. You can have the roof finished by a local powder-coater for a durable, long-lasting finish. We like rust, so we're going to let our sheet metal go natural. Standard roofing materials can also be used, though you might want a plywood underlay.

To install the roofing, center the sheet on the top of the coop and drill a couple of pilot holes through the steel and into the rafters. Fasten the roof with 2" roofing screws to secure it in place, then add eight more screws, evenly spaced on the rafters and driven through pilot holes.

10. COMPLETE THE FINISHING DETAILS.

Construct a simple nesting box using leftover plywood and short pieces of 2×2 for corner supports, as shown in *Nesting box and roosting bar*. You can size the box as desired, but the inside dimensions should be at least 12" w × 14" h × 12" d (see pages 15 and 17 for tips on planning roosting bars and nesting boxes). Assemble the box with 1¼" screws. We left our nesting box loose in the coop so we could move it around or remove it for cleaning as needed.

Create two notched cleats for a roosting bar, using leftover 2×4 lumber, as shown in *Roosting bar cleat detail* (see page 32 for help with cutting notches). Install the cleats on two opposing walls inside the coop; we placed ours on the side walls so the roosting bar sits on the nesting box to help support the weight. Center each cleat over a wall stud and fasten it with two 2½" screws driven through pilot holes and into the stud. Cut a 2×2 roosting bar to span across the coop and rest in the cleats' notches, and set the bar in place.

Finish the coop exterior as desired. Install latches and door handles, as needed (or other hardware; see page 27 for ideas) to the coop door and window.

Note: For our nesting box, we used ¾" plywood with rabbet joints at the corners. See page 17 for building tips.

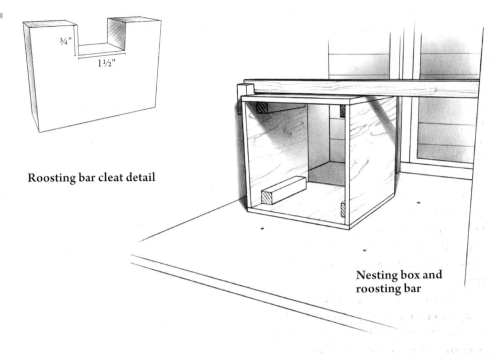

Roosting bar cleat detail

Nesting box and roosting bar

STOOP COOP

Is it possible for a chicken coop to be a gathering place? Instead of a utility structure in exile, what if a coop could be a place to commune, hang out, and even sit on? These were some of the playful questions behind the design of the Stoop Coop. We were well familiar with chicken coops as models of architectural expression; now we wanted to expand them into the realm of furniture. As a destination for informal hangouts, the Stoop Coop asks for engagement.

Stoop Coop was part of a project commissioned by City Slicker Farms, a nonprofit urban farming organization in Oakland, California. The coops were built in a grant-funded workshop and distributed to families free of charge. Upon delivery of the finished coops, a chuckle and a smile were the typical first reactions.

To make this design work, some modifications to a traditional staircase plan were called for. Each of the first three treads is angled slightly downward to shed water. The top tread is also sloped, but in the opposite direction, providing the ideal angle for a comfortable bench. The first and third treads open upward to give caretakers access for egg collection and replenishing food and water. For easy cleaning, there's a large door in the back. The first two "step" risers are meshed, providing shaded ventilation, and there is an additional mesh-covered window on one side.

We built our coops with marine plywood and added a tough clear-coat finish for durability out in the elements, but we know this won't stand up as well as a roofing material. If your coop shows signs of wear after a few years, try covering the treads with pieces of galvanized sheet metal or simply replacing the treads with new plywood. As an adaptation of this free-standing Stoop Coop design, you might build one under an existing staircase that leads from a deck or outbuilding.

> Three 4 × 8-foot sheets ¾" marine plywood
> Seven 8-foot 2×2s
> Five 8-foot rot-resistant 2×2s
> Five 8-foot 1×2s
> One hundred fifty 1½" deck screws
> Fifty 1¼" deck screws

> Fifty 2½" deck screws
> Four ¾" × 1½" exterior hinges with screws
> Two 1½" × 2½" exterior hinges with screws
> Exterior wood glue

> 12 square feet ½" galvanized hardware cloth, 18-gauge
> Finish materials (see step 6)
> Two spring-loaded exterior hasps with screws
> One exterior barrel lock with screws
> One exterior door handle with screws

1. CUT THE PLYWOOD SIDES.

Mark the layout of one of the coop sides onto one of the plywood sheets, using the short (4-foot) factory edge for the bottom of the piece. Cut out the side using a jigsaw or, preferably, using a circular saw for the bulk of the cutting, then finishing the corners with a jigsaw or handsaw.

Use the cut side as a template to trace cutting lines for the second coop side on the same piece of plywood, aligning the bottom edge of the "template" with the remaining short factory edge of the plywood panel. Cut out the second side. Save the cut-off material for the nesting box (step 3).

Cut out the window opening in one of the sides, using a jigsaw (starting with a drilled starter hole) or plunge-cutting with a circular saw (see page 34).

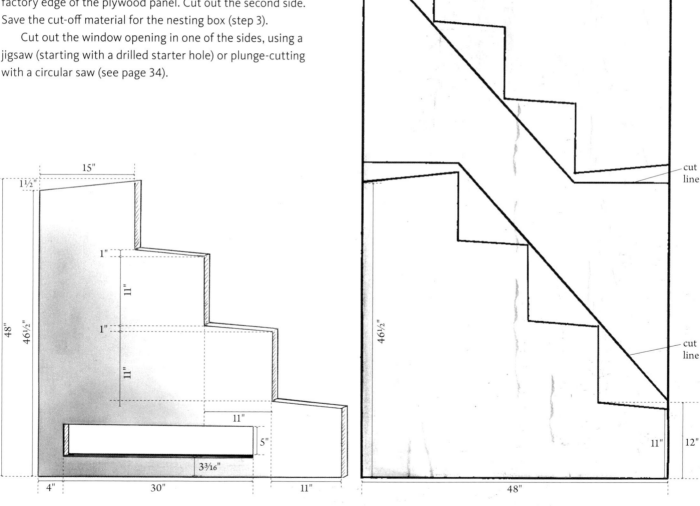

Side cutting diagram

2. ASSEMBLE THE COOP FRAME.

Cut the following frame pieces to length from 2×2 lumber:

- eight common pieces at 44½" for the step tread and riser supports
- two common pieces at 43½" for the vertical rear supports
- two rot-resistant pieces at 44½" and two at 45" for the floor supports
- two rot-resistant pieces at 45" for the skids on the underside of the coop

Assemble the coop frame by screwing through the plywood sides and into the ends or faces of the 2×2s, using 1½" screws. Position the tread supports at the front and back of each tread cutout, keeping the top faces of the lumber flush with the tread (horizontal) cut; this leaves a slight gap along the riser (vertical) cut. Drive two screws into each end of the tread supports.

Install the rot-resistant floor supports flush with the bottom edges of the plywood sides. Turn the coop on its side and install the rot-resistant skids directly underneath the side floor supports, using 2½" screws; these keep the rest of the coop 1½" off the ground for improved rot-resistance.

Cut and install the plywood floor to size at 48" × 44½". The floor will sit inside the coop and rest on top of the floor supports, so measure the interior space of your coop frame to confirm the correct dimensions before cutting. Notch the two rear corners of the floor panel to fit around the vertical 2×2s of the coop frame. Set the floor into position inside the coop.

44½"

45"

Assembled coop frame

3. BUILD THE NESTING BOX AND ROOSTING BAR.

Cut two pieces of plywood to size at 12" × 12" for the bottom and back wall of the nesting box. Join the pieces at a right angle, using wood glue and 1½" screws. Cut one common 2×2 to length at 44½" and one at 10½"; these are the supports for the nesting box. Add a 1×2 ledger to the lip of the nesting box.

Install the long support between the coop sides, as you did with the step tread supports, so that the nesting box will be accessible from the third tread of the completed coop. Fasten the short support to one of the coop sides with 2½" screws perpendicular to and level with the long support. Install the nesting box on top of the supports with 1½" screws.

The roosting bar is supported by notched cleats so that it's removable.

Cut two 2×2 cleats to length at 7½" each. Using a table saw or a circular saw and chisel, cut a 1½"-wide × ¾"-deep notch in the top center of each cleat. Cut the roosting bar from a common 2×2 at about 44¼" long, or just a bit shorter than the span across the coop interior. Install the cleats to the coop sides from the outside, using 1½" screws, so their bottom edges are 15¼" above the coop floor (make sure you have enough room below the nesting box support for easy removal of the bar).

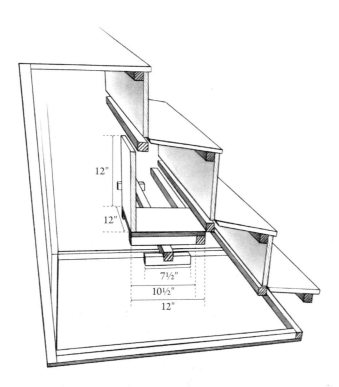

12"

12"

7½"

10½"

12"

Exploded view of nesting box and roosting bar

4. INSTALL THE STEP TREADS AND RISERS.

Cut the treads and risers from plywood. The treads for the first three steps are 12" × 48". The tread for the top step is 21" × 48"; the extra depth creates a 4" drip edge to keep water away from the door. Cut the two riser panels to size at 10¼" × 48"; these are for the top two risers. The bottom two risers are enclosed with hardware cloth. Cut two of the 12" treads from leftover plywood, and cut the remaining pieces from the third (uncut) sheet.

The first (bottom) and third treads are installed with hinges to provide access for egg collection and cleaning. Bevel the rear edges of two of the tread panels at 45 degrees, as shown, to provide clearance that prevents binding when opening and closing the tread "door."

Install the top three treads and top two risers, using 1½" screws. For the third-step (hinged) tread, install the riser first, then mount the tread to the riser with two ¾" × 1½" hinges. You will install the bottom-step tread after the first two risers are screened in with hardware cloth.

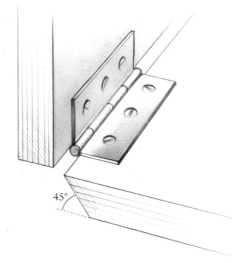

Tread bevel detail

5. CREATE THE BACK PANEL AND DOOR.

Measure and cut the plywood back panel to fit between the coop sides and extend from the top tread to the bottom of the rear floor support. Cut the door opening at 38" wide × 42" tall, giving the back panel an L shape. Install the back panel to the coop frame with 1½" screws.

Cut the door from the cutoff piece of the door opening, trimming the edges so the finished piece is ½" narrower and shorter than the opening. Cut a common 2×2 to run along the vertical edge of the door opening and install it to the interior side of the back panel; this provides backing for fastening the door hinges.

Hang the door with two 1½" × 2½" hinges, making sure there's an even gap along the top and side edges of the door opening. Cut two pieces of 1×2 to run along the top and latch-side edges of the door; these will serve both as stops and decorative trim. Fasten the stops so they overhang the door's edges by ¾", using 1¼" screws.

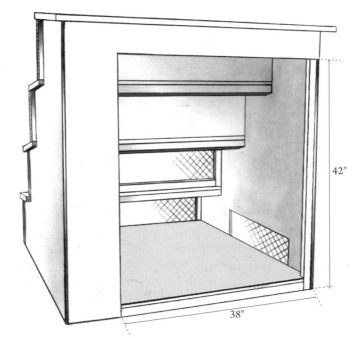

Back panel with door opening

42"

38"

6. INSTALL THE MESH AND FINISH THE COOP.

Cut pieces of hardware cloth to enclose the riser portions of the first and second steps on the front of the coop and to cover the window opening on the side. Secure the mesh with 1×2s and 1¼" screws (see page 16), installing the strips on the outside of the coop. Then, add 1×2s to support the back side of the mesh on the risers, spanning from side to side across the top and bottom edges of each mesh panel. Install the hinged tread for the first step, mounting the remaining two ¾" × 1½" hinges to the sandwiched 1×2s of the second riser.

Finish the coop by painting any accent details, as desired, then adding multiple coats of a clear protective finish.

Install a spring-loaded hasp latch on the hinged step treads, and add a barrel-bolt latch (or two, if desired) and handle to the back door.

MODERN MOBILE COOP-TRACTOR

GUEST DESIGNER: NICOLE STARNES TAYLOR OF MAKE DESIGN STUDIO IN SEATTLE, WA

This coop is ideal for urban lots where space is tight. The designer and her husband live in Seattle. They liked the idea of having chickens for composting their food, fertilizing plants, working the soil for new garden beds, and tilling under their vegetable garden at the end of the year. The eggs, which are creamy and delicious, are simply a bonus for them.

The size of the chicken coop was determined by the width of their home's "planting strip," Seattle's planted paradise between the street and the sidewalk. Their first project for the chickens was to convert the gravel- and weed-filled planting strip into a rich soil ready for new plants. This required a coop that could convert into a chicken tractor to allow the chickens to work the ground directly; all this takes is removing the screws between the coop house and its rollable floor (a.k.a. the "dolly") and setting the house structure directly on the ground. The coop also needed to be sturdy enough to keep the chickens safe from dogs and raccoons.

While sources on raising chickens typically recommend a minimum of three birds (and this coop is large enough for three), the designers found in their six years of raising chickens in an urban environment that two birds are ideal. With two, the chickens are quieter and more content, and they don't have the same dominance issues common among larger flocks.

One unexpected result of this coop's mobile design is how fun and easy it is to share the experience of raising chickens with others. When the designers started a major home construction project, they rolled the coop and chickens down the street to a friend's house. The friends were interested in raising chickens but had some reservations. The "loaner chickens" and coop gave them the opportunity to try it out without taking the plunge. One month into their chicken-sitting, the friends started building their own coop and had chickens of their own before they returned the loaners.

MATERIALS

- Two 4 × 8-foot sheets ½" CDX plywood
- Eight 8-foot 2×4s (common lumber, or use rot-resistant cedar for greater weather resistance)
- Four 8-foot 2×2s (same as above)
- One 4 × 8-foot sheet ¾" CDX plywood
- Nine 5-foot rough-sawn cedar 1×3 fencing boards
- One hundred 3" deck screws
- One hundred 1½" deck screws
- Fifty 2½" deck screws
- Six 6" deck screws
- Heavy-duty staples or 1" exterior screws with washers (see step 8)
- One hundred fifty 1½" exterior trim-head screws
- Two 6" swiveling, locking casters (mounting plate must fit 2×4 face; see step 1)
- Two 6" fixed casters (nonswiveling, nonlocking; same as above)
- One-half gallon exterior primer
- One-half gallon exterior paint
- 20 square feet ½" galvanized hardware cloth, 18-gauge
- 16 square feet expanded metal that is smooth on both sides (see step 8)
- 18½ square feet roofing material (see step 9)
- Two exterior 3" hinges with screws
- Two exterior 2" hinges with screws
- Two exterior door pulls with screws
- Three exterior barrel-bolt latches with screws
- One branch (or 2×2; for roosting bar)

SPECIALTY TOOL

- Grinder with metal-cutting wheel or circular saw with metal cutoff disc

OPTIONAL SPECIALTY TOOLS

- Miter saw
- Table saw
- Pocket hole jig (and screws; see step 2)
- Metal file
- Router

A few helpful hints for this project

- Use screws for all fastening, making cleaning and future disassembly much easier.

- Feel free to experiment and make material substitutions based on the simple form of the original coop design.

- Use recycled, found, or salvaged material as much as possible.

- Choose a solid floor over a screened floor. The designers have tried both, and the solid one is easy to clean with a flat shovel and leaves less feed on the ground, reducing night visits from rats and raccoons.

- Fill the wall space between the screen and the siding with straw in the winter, if desired, for cold weather protection.

- Build the coop before you get the chicks. The designers learned this the hard way. Baby chicks grow fast. Having chickens in the living room isn't nearly as fun as it sounds.

- You can replace the plywood roof with sheet metal or any other easily sourced roofing material.

1. BUILD THE COOP FLOOR.

Cut the floor piece to size at 32" × 60" from ½" CDX plywood, using a circular saw and a straightedge guide. Cut two 2×4s to length at 60", mitering both ends at 45 degrees in opposite directions. It's easiest to make these cuts with a miter saw, but you can use a circular saw or even a handsaw and a miter box. Measure the length of all angled cuts from the long points on the miters.

Cut two more 2×4s at 32" with opposing 45-degree miters. Arrange the 2×4s on top of the plywood to create a frame that aligns with the edges of the plywood. Fasten together the 2×4 frame at the corners, using 3" screws.

Place the frame on a flat surface and position the plywood floor on top of the frame so all outside edges are flush. Screw through the plywood and into the frame pieces with 1½" screws. Flip the assembly over so the 2×4s face up. Install four casters on the 2×4 frame, placing the two swiveling casters on one end of the frame and the two straight casters at the other end.

NOTE: *As an optional detail for more experienced woodworkers, you can cut a rabbet into the 2×4 frame and recess the plywood (cut smaller than specified above). This provides a cleaner look on the outside by hiding the plywood edge.*

2. CONSTRUCT THE SIDE WALLS.

Cut the following pieces for each of the two angled side walls of the coop (remember to measure from the long point of all angled cuts):

- One 2×4 at 60"
- One 2×4 at 25½", with a 9-degree angle on one end
- One 2×4 at 34¼", with a 9-degree angle on one end
- One 2×4 at 60¾", with parallel 9-degree angles on both ends
- One 2×2 at 30½", with a 9-degree angle on one end
- One 2×2 at 32½", with a 9-degree angle on one end

Following the *Side wall diagram*, assemble each wall frame using 2½" screws. If you have a pocket hole jig (and appropriate screws), you can fasten all of the joints from the inside, which makes for clean, easy work.

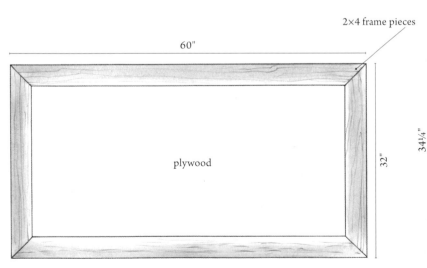

2×4 frame pieces

60"

32"

plywood

Coop floor — bottom view

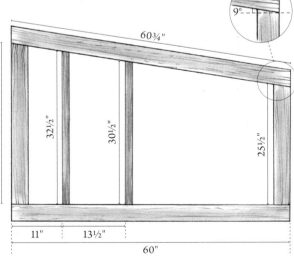

9°

60¾"

34¼"

32½"

30½"

25½"

11"

13½"

60"

Side wall diagram

3. PREPARE THE DOOR AND FRONT WALL PIECES.

You will assemble the frame structure of the coop after preparing the front wall (and door frame), the rear wall, and the roof. To prepare the front wall, cut two 2×4s to length at 29" to serve as the top and bottom rails of the front wall. Bevel the top edge of the top rail at 9 degrees to follow the slope of the angled side walls, using a table saw (if you have one) or a circular saw.

Cut two 2×2s at 34" and two at 28¾", all with opposing 45-degree miters at the ends. Assemble these pieces picture-frame-style, using 2½" screws to create the door frame (the frame will be much more stable after expanded metal is installed on its back side, in step 8).

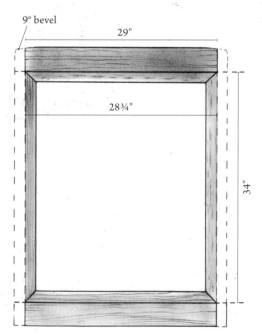

Front wall diagram

4. BUILD THE REAR WALL FRAME.

Cut two 2×4s to length at 29" for the rear top and bottom rails. Bevel the top edge of the top rail at 9 degrees to match the slope of the angled sides, as you did with the front wall's top rail.

Cut two 2×2s to length at 25". Assemble the rails with 2½" screws as shown.

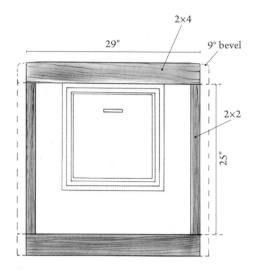

Rear wall diagram

5. CUT THE ROOF DECK.

Cut the roof deck to size at 36" × 72", using ½" plywood. Lay out and cut a handle at the front end of the roof deck, following the *Roof deck diagram*. Use a circular saw with a plunge-cutting technique (see page 34) or a jigsaw (by drilling a starter hole to insert the blade inside the cutout).

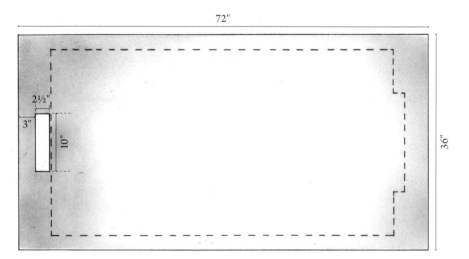

Roof deck diagram

6. ASSEMBLE THE COOP FRAME.

Stand up the angled side walls on top of the coop floor, and position the 2×4 rails of the front wall between the sides. Join the walls by screwing through the sides and into the rails with 3" screws. Do the same with the rear-wall rails, using 2½" screws. Make sure the wall assembly is aligned with the edges of the floor, then fasten the walls to the floor with 6" screws driven down through the top edges of the walls' 2×4s. The swiveling casters should be at the front end of the coop.

Position the roof deck on top of the wall assembly, following the *Roof deck diagram* in the previous step. Fasten the deck to the wall frames with 1½" screws. Confirm that all fits well, then remove the roof for the next two steps.

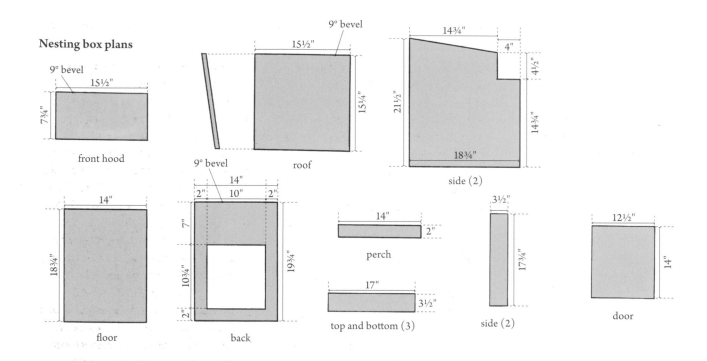

Nesting box plans

front hood — 9° bevel, 15½", 7¾"

roof — 9° bevel, 15½", 15¼"

side (2) — 14¾", 4", 4½", 21½", 14¾", 18¾"

floor — 14", 18¾"

back — 9° bevel, 14", 2", 10", 2", 7", 10¾", 2", 19¾"

perch — 14", 2"

top and bottom (3) — 17", 3½"

side (2) — 3½", 17¾"

door — 12½", 14"

7. BUILD THE NESTING BOX.

Lay out and cut all of the nesting box parts from ¾" plywood, following the *Nesting box plans* (previous page). Assemble the box with 1½" screws, but leave off the door for now. You will hang the door and install the nesting box onto the coop in step 9.

Nesting box side view

Nesting box front view

Nesting box back view

8. PAINT AND SCREEN THE COOP.

Disassemble the coop, then prime and paint all of the wood parts. The designers used Glidden Exterior Latex Satin GL6912 (Orange Marmalade) for a bright, yolk-yellow effect. Be particularly thorough in painting the end grain of the wood to protect against moisture.

NOTE: *Do not paint the top face of the coop floor or the nesting box floor, so there's no risk of the chickens' scratching loose and eating paint chips.*

After the final coat of paint has fully dried, cut and install hardware cloth and expanded metal on the insides of the walls and front door. Each of the side walls gets hardware cloth on the two inner sections, which will be covered on the outside by siding slats (step 10). All other openings and the front door get expanded metal (see note at the end of this step). Cut the hardware cloth and expanded metal so it overlaps the frame parts by at least 1", and fasten it with heavy-duty staples or 1" exterior screws with washers.

Install the hardware cloth first. At the rear corners of the coop, on the side walls, position the hardware cloth 2" back from the rear edge to allow the rear wall 2×2s to fit snugly against the side walls. Cut the expanded metal with a grinder, a circular saw, or a jigsaw with a metal-cutting blade. Install the expanded metal to the front door frame and the side walls.

NOTE: *Be sure to use expanded metal that is smooth on both sides. Do not use stucco lath or similar materials that have a rough texture or any sharpness on either side.*

9. REASSEMBLE THE COOP AND COMPLETE THE ROOF.

Reassemble the coop by screwing the parts back together just as they were before. Install the roof deck. Install the nesting box next, before screening in the rear wall: Position the box against the roof and top rear-wall rail and fasten it to both elements with screws (it's good to have a helper for this). Notch the expanded metal screen for the rear wall to fit snugly around the sides and bottom of the nesting box. Position the screening so the notched edges are behind the plywood trim of the nesting box, and fasten the screening to the rear-wall framing and the box's trim.

Mount a door handle to the outside of the nesting box door (the beveled top edge faces the inside of the box). Hang the door, using two 2" hinges screwed into the nesting box floor and the inside face of the door. Make sure the door doesn't rub the box floor; you may need a 1⁄16" shim under each hinge to raise the door slightly. Also make sure the door is centered side-to-side in the box opening.

Cover the roof deck with any roofing material you like: You can use corrugated metal (see page 26), standard shingles, or something creative, like reclaimed street signs or shingles cut from aluminum cans. The designer used hot rolled steel.

10. INSTALL THE SIDING SLATS.

Cut 16 pieces of rough-sawn cedar 1×3 fencing boards to length at 35¾" and cut 18 pieces to length at 12¾". These are the siding slats for the coop's side walls. At this point, you have the option of finishing the slats with stain. The designer did not stain the slats. If you choose to do so, use a pigmented stain to prevent the wood from turning gray over time. Install the slats on the outside of the side walls, as shown in the *Siding layout*. Fasten the siding to the wall framing, using two 1½" exterior trim-head screws at each joint.

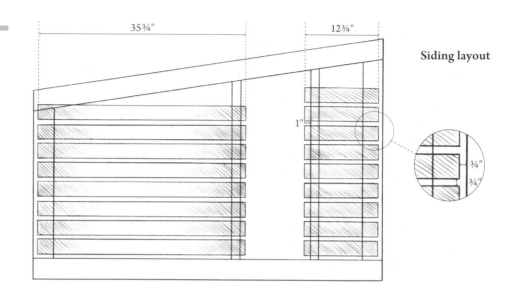

Siding layout

35¾"

12¾"

1"

¾"

¾"

11. COMPLETE THE COOP.

Hang the front door with two 3" hinges mounted to one of the side-wall frames. Install a pull on the front door, and add barrel-bolt latches to both doors. The front door gets one at the top and one at the bottom; the nesting box door should have at least one at the top, but you can use two for additional security.

Cut the roosting bar to fit snugly between the 2×2s located closest to the nesting box. The designers used a stout branch for their roost, but a 2×2 will work, too. Install the roosting bar about 23" from the coop floor and about 18" from the front of the nesting box, using screws.

TIP: *To add a waterer and feeder, you can hang them from the roof, install a rod across the width of the coop, or build a shelf.*

Note

The designer made a few improvements to the coop that are not shown in the photos but are reflected in the text and drawings. The hand hold shown in the roof over the nesting box was omitted, and two 2×2s were added on the rear panel to provide a better fastening surface for the expanded metal.

ICEBOX

In the summer of 2010, we built two Icebox coops with City Slicker Farms in West Oakland, California. The coops now reside in backyards in the neighborhood, providing fresh eggs for a few lucky residents.

The Icebox is named for its resemblance to a deep freezer. It was designed to be light enough to move around the garden and to be easy to clean. This is a small coop, intended just for locking up the hens at night. Its minimalist design makes it easy to assemble in addition to being easy on the wallet. It's also low to the ground, which makes it great for children to be able to participate in the chicken-raising process. If you build one, you should also have a bigger yard or run for the chickens to live in during the day.

The Icebox has many plywood surfaces to paint, so don't be afraid to get creative with color and design. We used clear plastic corrugated roofing material to let more light into the coop. Without much effort, you could add windows or doors on the sides or back of the coop. The shape would scale up nicely: Double the size to build a big coop that could easily house eight chickens. Or, mount it on 4×4 stilts to raise the coop up to a higher level for access and cleaning.

MATERIALS

- Two 4 × 8-foot sheets ¾" plywood
- One 8-foot rot-resistant 2×4
- Four 8-foot 2×4s
- Two 8-foot rot-resistant 2×2s
- Three 8-foot rot-resistant 1×2s
- Five 8-foot 2×2s
- One 8-foot 1×2
- Fifteen 3½" deck screws
- Twenty-five 2½" deck screws
- One hundred fifty 2" deck screws
- One hundred 1¼" deck screws

- Thirty 1½" roofing screws with neoprene washers
- 8 linear feet ½" galvanized hardware cloth, 18-gauge, 36" wide
- Four 2½" metal L straps
- Two 3" metal T straps
- 24 square feet metal or clear plastic corrugated roofing
- Three 3½" galvanized hinges with screws
- One quart exterior paint

- One half-gallon clear exterior finish
- One 4" exterior door handle with screws
- Two exterior hook-and-eye latches or hasp latches with screws

SPECIALTY TOOL

- Table saw

1. CUT THE PLYWOOD SIDES.

Draw out one of the plywood sides onto one sheet of ¾" plywood, and make the cuts with a circular saw or jigsaw. Use the cut piece as a template to trace the other side onto the remaining plywood so both sides are identical. You can use the opposing corner of the same plywood sheet to simplify the cutting. Cut out the second side piece. Save the plywood cutoffs for other parts of the coop.

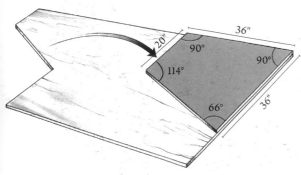

Cutting side pieces

2. COMPLETE THE SIDE WALLS.

The plywood sides get a 2×4 frame on their inside faces to complete the structural side walls of the coop. The easiest way to mark the cuts for the 2×4s is to scribe each piece using the plywood sides: Place a rot-resistant 2×4 along the bottom edge of a plywood side piece and trace along the plywood edges to mark the 2×4 so it will be flush with the sides of the plywood. Cut the 2×4 and reposition it at the bottom of the plywood again, and hold or clamp it in place.

Use the same technique to mark and cut the two vertical side pieces, using common 2×4 lumber and tracing along the top edge of the plywood. Scribe and cut the top piece to fit between the vertical 2×4s.

Install the 2×4s by drilling ⅛" pilot holes and screwing through the outside faces of the plywood and into the lumber pieces with 2" screws. Do not screw near the corners of the plywood at this stage. Keep in mind the two sides mirror each other, with plywood on the outside.

3. COMPLETE THE COOP FRAME.

Cut two common 2×4s and two rot-resistant 2×2s to length at 60"; these are the horizontal rails that run between the side walls. Use a table saw to bevel the top edge of each 2×4 top rail at 66 degrees.

Install the 2×2 rails between the bottom corners of the side walls, drilling pilot holes first, then using 3½" screws driven through the outsides of the side walls. Install the beveled top rails at the top corners of the walls in the same fashion.

Cut a rot-resistant 2×2 floor support to fit between the two bottom rails, and install it with one 2½" deck screw at each end. Cut a 2×4 door post to fit between the front bottom rail and front top rail; install this with 2" screws toe-screwed (see page 29) through each side edge of the post and into the top and bottom rails.

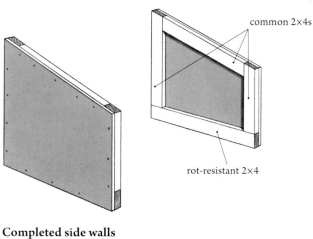

common 2×4s

rot-resistant 2×4

Completed side walls

66°

60"

Top rail detail

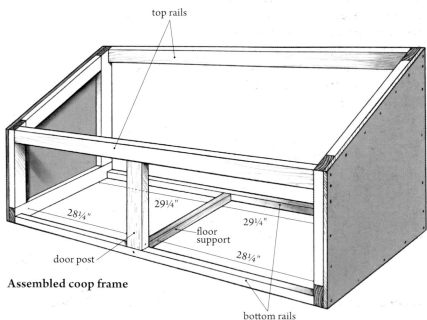

top rails

28¼"

29¼"

29¼"

floor support

door post

28¼"

bottom rails

Assembled coop frame

4. ADD THE FLOOR.

The underside of the coop is completely covered with ½" wire mesh to keep predators out. On top of the mesh, the left half of the coop has a plywood floor, which is removable for cleaning or replacement.

Lay out and cut the floor panel from plywood. Set the panel in place on top of the bottom rails and floor support so its right-side edge is flush with the right face of the floor support. Fasten the panel with six 1¼" screws.

Cut a piece of hardware cloth to size at 63" × 36" (see page 26). Cut two rot-resistant 1×2s to length at 63" and three at 33". Cover the bottom of the coop with the mesh, securing it with the 1×2s and 2" screws.

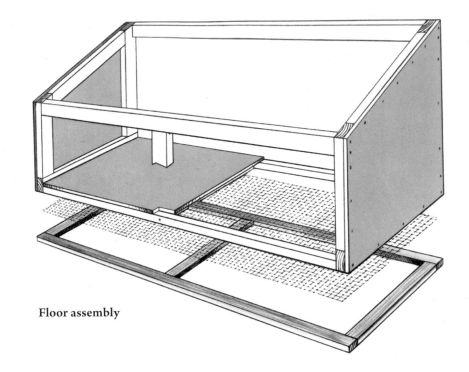

Floor assembly

5. BUILD THE ROOF.

Cut common 2×2s for the roof frame: two front and rear pieces at 63", one crosspiece at 60", and two side pieces at 36". Assemble the frame with one 2½" deck screw at each joint. If you'd like to paint the roof frame like we did, it's easiest to do it now.

Check the completed frame for square, then reinforce each joint with L-straps (on the four corners) and T-straps (for the crosspiece), using 1¼" screws. Mount the roof frame to the rear top rail of the coop with three 3½" hinges.

Cut corrugated roofing panels to length at 45". This length gives you 3" of overhang at the front and rear of the coop. You can make the panels longer for deeper overhangs, but you want to make sure to have easy access to the handle on the front of the roof frame (see step 9). Install the roofing panels with roofing screws with neoprene washers (see page 26 for cutting and installation tips).

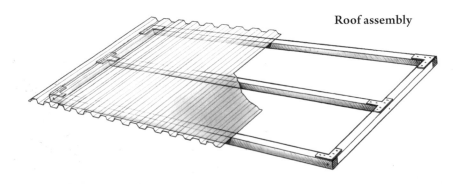

Roof assembly

6. ENCLOSE THE FRONT AND REAR.

Cut a plywood back panel to size at 63" × 36". Install the back so it's flush with the tops of the coop sides, using 2" screws driven through the back and into the sidewall 2×4s and the top and bottom rails.

Cut a piece of hardware cloth for the right half of the coop's front, at 31" × 17½". Then, cut 1×2s for securing the mesh: one rot-resistant piece at 31½", one common piece at 31½", and two common pieces at 15½". Install the mesh over the right opening at the front of the coop, securing it with the 1×2s and 2" screws; use the rot-resistant piece at the bottom (we painted our 1×2s at the end of the project).

Cut plywood for the front door at 31½" × 17½". Hang the door with two 2½" hinges mounted to the front bottom rail of the coop and the bottom of the door.

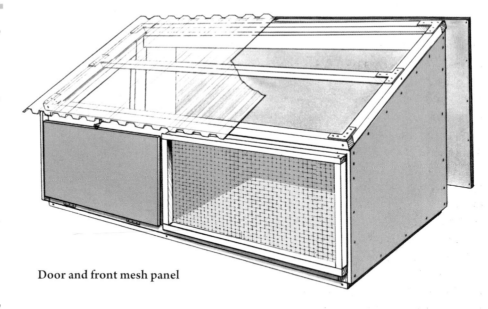

Door and front mesh panel

7. ADD THE ROOSTING BAR.

Cut a common 2×2 to length at 63" for the roosting bar. Install the bar with one 2" screw on each side, driven from the outside of the coop. The bar also serves as a support for the nesting boxes.

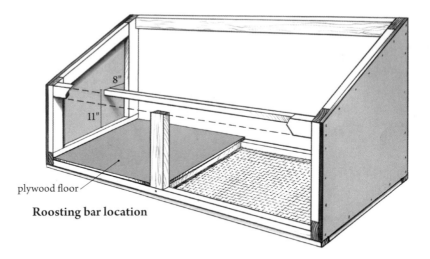

Roosting bar location

8. BUILD THE NESTING BOXES.

Construct the nesting boxes from the remaining plywood: Cut a shelf to size at 36" × 18", then notch one corner to fit around the vertical side-wall 2×4 at the right rear corner of the coop. Position the shelf on top of the roosting bar, and make sure it's level from front to back. Scribe along the top edges of the shelf with a pencil to mark the shelf location onto the back and side panels of the coop. Remove the shelf, then drill ⅛" pilot holes through the side and back panels ⅜" under the scribed lines. Set the shelf back in place and fasten it to the roosting bar with 2" screws. Then drive 1¼" screws through the back and side panels and into the edges of the shelf.

Cut two dividers from the leftover plywood. You can modify the dimensions as desired, but make sure the roof won't hit the dividers when it closes. Position each divider on the shelf, then scribe along the back panel and drill pilot holes as you did with the shelf. Fasten the dividers to the back panel with 1¼" screws. It helps to have a friend hold the dividers while you drive the screws from the back of the coop.

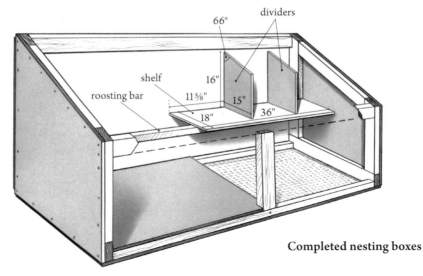

Completed nesting boxes

9. ADD THE FINISHING TOUCHES.

Finish the coop with paint or another exterior finish as desired, and apply two layers of clear coat.

Install an exterior door handle centered along the front of the roof frame.

Install two hook-and-eye latches or hasps to secure the roof frame to the door and the mesh section at the front of the coop.

The Humble 2×4

If you've tackled many building projects around the house, you may have asked yourself, "Why does a 2×4 actually measure 1½" × 3½"? Are we getting ripped off here?" It turns out the answer is not that simple.

Until the mid-1800s, lumber was typically produced locally, and suppliers catered to the specific needs of local carpenters. But as forests became depleted around cities, lumber had to be shipped increasingly greater distances, and there were no standards of lumber sizing. In order to compete with local mills, lumber shipped from farther away was cut slightly smaller to reduce shipping costs. At a lumberyard in 1910, one might find a 2×4 from any number of suppliers, all in different dimensions. Also, some manufacturers shipped wet lumber (with a high moisture content) at 2" × 4", which then shrunk considerably as it dried. The opening of the Panama Canal in 1914 further confused things, as western lumber was more easily and cheaply shipped to eastern ports.

Thankfully, in the 1920s and 1930s various manufacturers, shippers, associations, and committees began to develop standards for the lumber industry. The difference between the nominal size (2" × 4") and actual size (1½" × 3½") has been standardized across the country. Similarly, lumber in many common nominal sizes (1×2, 2×2, 2×4, 4×4, and so on) became available consistently throughout the United States. In addition, restrictions were placed on the number of knots, the grade of lumber, the grain, and the moisture content. These standards helped fuel the post–World War II building boom and allowed us to cheaply and easily purchase lumber to build the Modern Log Cabin chicken coop.

Today, the 2×4s used in construction are usually composed of plentiful (and therefore inexpensive) softwoods such as Douglas fir, pine, or spruce. The wood used in these 2×4s is from the outer layer of the tree, just under the bark, which is strong enough for standard framing lumber and more plentiful than the inner, denser, and stronger part of the tree, called the heartwood. Heartwood is commonly used for structural-grade lumber (for beams and other critical load-bearing members) and for premium grades of decking. The 2×4 is a strong and versatile piece of wood, available in lengths from 4 feet to 24 feet, and used in framing, trussing, and bracing.

We here on America's West Coast are blessed with sustainably managed forests, cheap and plentiful lumber, and consistent lumber products. The Forest Stewardship Council (FSC) is a third-party certification organization that labels products as sustainably harvested according to strict environmental guidelines. We encourage you to look for their logo (below) on any wood products you purchase. We are conscious that trees are a limited resource, so we always take care to reduce our waste and to buy reclaimed lumber whenever possible. Even with the standards that have developed over the last 100 years, there remain great differences in quality between two pieces of wood. When choosing your 2×4s, pick up one end of each piece and sight down one edge (like looking down the barrel of a rifle). Choose lumber that is as straight as possible, preferably not bowed or twisted.

FSC   The Forest Stewardship Council logo

MODERN LOG CABIN

We had an idea to build a log cabin chicken coop. Log cabins are so old, so American, so frontier, so homesteader. One of our primary goals with this book is for the reader to be able to build our coops with readily available materials, and procuring a bunch of logs isn't so easy for urban dwellers. So we came up with a design that features a common modern building material — the humble 2×4 — while using log-cabin joinery for assembly. Essentially, we're building an old-style structure with a modern twist.

While 2×4 lumber is cheap and readily available all over the United States, you can also find used 2×4s from home remodels, at scrap yards and building materials recyclers, and through Web resources and word-of-mouth.

This coop was really fun to build. It took us a few hours to cut and notch all of the 2×4s, but after that the coop went up fairly quickly. It is *very* important to make accurate measurements and cuts with the 2×4s, as slight differences in length will make your coop difficult to put together.

Our coop features a sliding barn door, an asymmetrical pitched roof, and a clear plastic roof panel for admitting sunlight, but the possibilities for custom doors, windows, and roof styles are endless.

Keep in mind this cabin will start to get heavy as you build upwards. We built ours in the shop, then took the roof structure off to carry the building outside. We installed the roof once the coop was in place. We strongly recommend that you assemble this coop on-site, if possible, so you won't have to move it into position.

MATERIALS

- Thirty 10-foot 2×4s
- Two 8-foot 2×4s
- One 10-foot rot-resistant 2×4
- One 10-foot rot-resistant 2×6
- One 10-foot 2×6
- Twelve 6-foot 2×2s
- Nine 10-foot 1×2s
- One 8-foot 2×4 (for homemade door track; see step 11)
- One 4 × 8-foot sheet ¾" plywood
- Two hundred 2" deck screws
- Twenty-five 1¼" deck screws
- Ten 2½" deck screws

- Fifteen 3" deck screws
- Seventy-five 1½" roofing screws with neoprene washers
- 10 linear feet ½" galvanized hardware cloth, 18-gauge, 36" wide
- 40 square feet metal and/or clear plastic corrugated roofing
- Four fixed (not swiveling) casters (for homemade door track; see step 11)
- 10 linear feet of wiggle board
- 6-foot length 4" × 4" galvanized flashing

- Two 2" exterior hasp latches with screws
- One pint exterior red paint (or the color of your choice)
- ½ gallon natural deck sealant

OPTIONAL SPECIALTY TOOLS

- 1" Forstner drill bit
- Bandsaw
- Dead-blow mallet or rubber mallet
- Jigsaw
- Table saw with dado blade

---

1. CUT THE LUMBER TO LENGTH.

To ensure accuracy and consistency, use a stop block setup with a miter saw to make the square cuts on the lumber that will become the essential building blocks of your cabin. Make sure the stop is square to the saw blade, and check your cut lengths frequently for accuracy. Sometimes boards from the lumber yard don't have square edges, so we cut a little off one end and measure from there.

Square off one end of 21 of the thirty 10-foot 2×4s, then cut 42 pieces to length at 57".

Square off and cut the following additional pieces:

- Two rot-resistant 2×4s at 57"
- Two rot-resistant 2×6s at 57" (see note)
- Two common 2×6s at 57" (see note)

- Six 2×2s at 67"
- One 2×2 at 48"
- Two 2×4s at 71"

NOTE: *There may be a difference in the actual widths between the common and rot-resistant 2×6s. If so, you may need to rip along the entire length of some of the pieces*

to end up with 5¼" in total width. Our rot-resistant 2×6s came from the lumberyard at 5½" wide, while the common 2×6s were 5¼". So we ripped ¼" from the rot-resistant pieces to match the width of the commons. It is important to rip the pieces before you notch them in the next step.

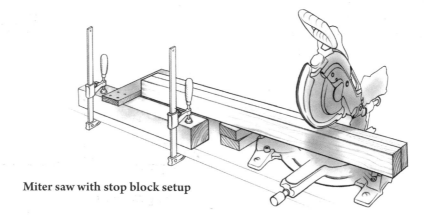

Miter saw with stop block setup

2. NOTCH THE LUMBER.

The best tool for notching the pieces is a table saw with a dado blade (see page 32), but you can also use a miter saw with a depth stop or a circular saw and chisel. Note that both ends of each piece should be notched.

Set up your saw to cut the notches in the 57" 2×4s, as shown in the *Notching diagrams*. Use two pieces of scrap 2×4 to cut and test-fit your first notch. Adjust the depth and width of the cut as needed so the top and bottom of the mating 2×4 pieces are flush. If necessary, use a mallet to gently tap the pieces together, but you shouldn't have to apply much pressure. The pieces should fit together snugly but easily.

Notch all 42 of the 57" 2×4s. The two 57" rot-resistant 2×4s do *not* get notched. With the same blade setting, notch the two 57" 2×6s and the two 71" 2×4s as shown in the notching diagrams.

Reset the depth of the blade and notch the six 67" 2×2s (these are roof rafters). The 48" 2×2 is the apex rafter and does *not* get notched.

Reset the depth of the blade again and notch the two 57" rot-resistant 2×6s.

NOTE: *It's a good idea to test-fit each notch after resetting the blade, especially if you're using expensive lumber such as cedar or redwood.*

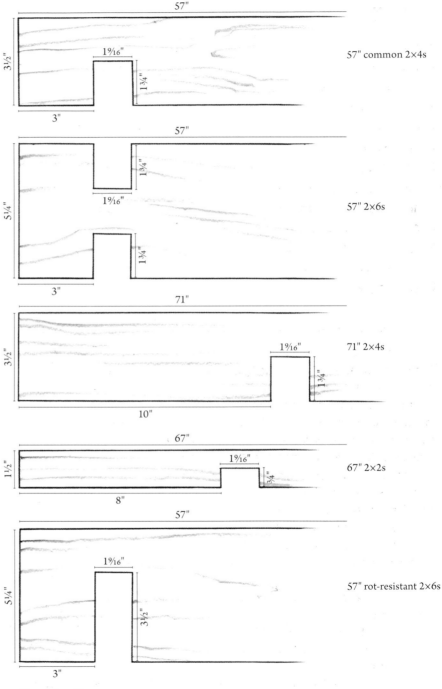

Notching diagrams

3. BUILD UP THE WALLS.

Position the two rot-resistant 2×4s (which do not have notches) on edge on a flat surface. Set the two rot-resistant 2×6s — with notches facing down — onto the 2×4s to create a square. Check carefully to make sure the assembly is square. Predrill from above with a ⅛" bit, and drive one 2" screw per joint through the 2×6 and into the 2×4.

Because the screws will go right on top of each other, we found it helpful to stagger the screws between layers: a screw on the inside half of one layer, then a screw on the outside half of the layer above.

Begin working your way up with common 2×4s. Predrill and screw as before, one screw per joint, per layer, alternating inside and outside. Check to make sure the corners are square and the walls are plumb as you go. The walls that started with the 2×6s get a total of 10 common 2×4s and finish with one common 2×6 on top. The other two walls get 11 common 2×4s and will finish with the placement of the roof structure.

4. CUT THE BARN DOOR OPENING.

On one of the walls that started with a rot-resistant 2×4, mark and cut the barn door opening at 21½" wide × 28" tall. Make the cuts by plunge-cutting with a circular saw, then finishing each cut with a handsaw, being careful not to cut into the adjoining 2×4. Wrap the inside edges of the opening with common 1×2s, cutting each piece to fit and fastening it with 2" screws.

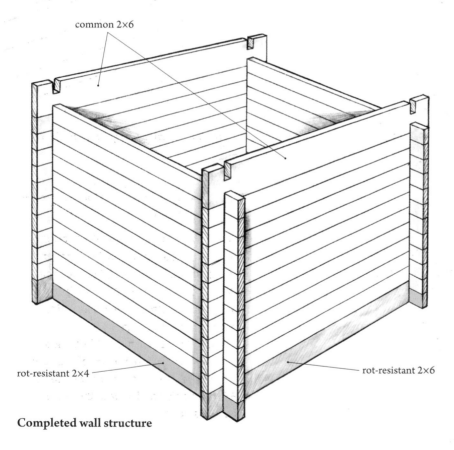

common 2×6

rot-resistant 2×4

rot-resistant 2×6

Completed wall structure

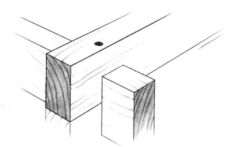

Screw detail, staggered outside

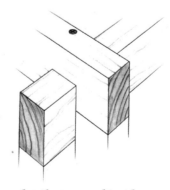

Screw detail, staggered inside

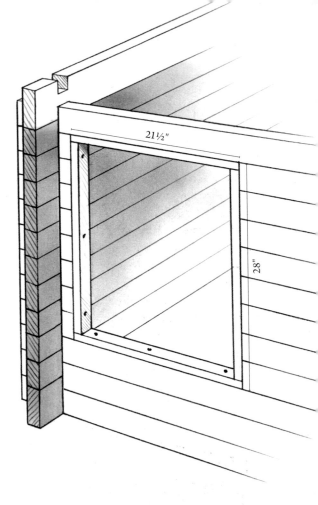

Barn door opening wrapped with 1×2s

21½"

2.8"

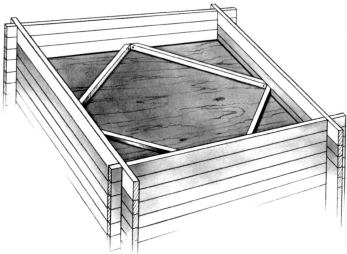

Bottom view of cabin with floor and ledgers installed

5. INSTALL THE FLOOR.

The cabin floor is a square of ¾" plywood that's supported below by 2×2 ledgers fastened to the 2×4 walls. Cut the cabin floor to size at 47½" × 47½". Cut four 2×2 ledgers to length at 32" each, mitering both ends at 45 degrees.

Install the ledgers with 2" screws so they are level, and so the top of the floor will be flush with the barn door opening. Fasten the plywood floor to the ledgers with 2" screws.

Completed barn door

6. MAKE THE BARN DOOR.

Using a stop block setup on a miter saw, cut seven common 2×4s to length at 31". Cut three common 1×2s to brace the door in a classic Z formation. Arrange the 2×4s so their ends are perfectly even, and attach the 1×2 braces with 1¼" screws, driving one screw through the 1×2s and into each 2×4.

Cut two more common 1×2s to cover the top and bottom of the door, and install them with 1¼" screws.

If desired, paint the exterior of the door red, leaving the inside face natural or painting with a neutral color. Some animal experts believe chickens hate the color red, so don't paint the inside of your barn door red, as it might cause your chickens to be uneasy in their new home.

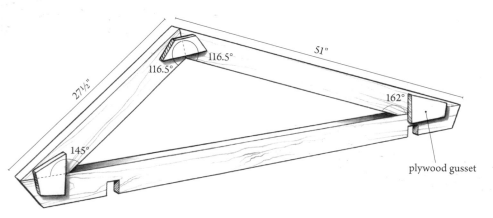

116.5°   116.5°   51"   162°
27½"   145°   plywood gusset

Gable truss diagram — interior view

7. FRAME THE GABLE TRUSSES.

The gable trusses are two triangular roof supports that define the shape of the roof and support the ends of the horizontal roof rafters. First you will frame the trusses, then you will cover them with hardware cloth, which provides light and ventilation for the coop while keeping out predators. The roof described here has an off-center peak, but your roof can have a more traditional gable shape if you prefer.

Starting with the two 71"-long 2×4s that you notched in step 2, bevel both ends of each piece at 30 degrees so the long points of the miters are on the edges that are *not* notched. Then, square-cut each of the two 8-foot 2×4s into two pieces roughly 35" and 61" long; cut these two pieces again using the angles shown here to form the truss triangle. You might find it easiest to fit the triangle together by overlapping the pieces and scribing and cutting the angles one at a time and/or using a sliding T-bevel (see page 30).

Dry-assemble each truss on a work surface and mark the rafter locations, spacing them roughly 16" apart. Note that the trusses will oppose each other on the coop, so mark the rafters accordingly. The 48" 2×2 apex rafter abuts the roof triangles, so it is not notched.

Using a bandsaw, jigsaw, or circular saw, cut six plywood gussets for joining the corners of the truss members, sizing the gussets so they won't interfere with the rafters or coop structure. Assemble the trusses by fastening a gusset at each joint, using 2" screws, locating the gussets on what will be the inside faces of the trusses.

8. COVER THE GABLE TRUSSES.

Using a sliding T-bevel, scribe and cut common 1×2s to run along the centers of the truss members. Cut the 1×2s to avoid interfering with the notches on the 2×4s. Position the 1×2s on the outside faces of the trusses (on the sides opposite to the plywood gussets) and temporarily fasten each piece with a couple of 2" screws. Lay each truss upside down on the hardware cloth, and with a permanent marker trace along the outer edge of the 1×2 triangle. Be sure to do this for each truss, as the 1×2 triangles may be slightly different.

Use shears to cut the two pieces of hardware cloth (see page 26). Notch out the hardware cloth inside the two bottom notches of each horizontal truss member. Unscrew the 1×2s from the trusses, lay in the hardware cloth, and sandwich the cloth between 1×2s and truss members, adding more screws for a permanent installation.

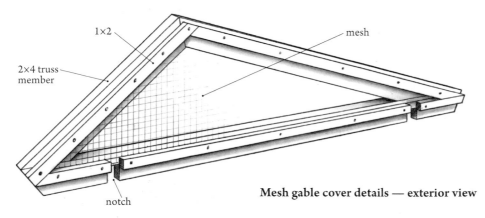

1×2

2×4 truss member

mesh

notch

Mesh gable cover details — exterior view

9. COMPLETE THE ROOF STRUCTURE.

Install each gable truss so the 1×2s are on the outside and the plywood gussets are on the inside. Fit the notches at the bottom of each truss onto the notched 2×6s of the wall structure, and toe-screw the trusses with 2½" screws.

Miter the ends of the six 67" 2×2 rafters (notched in step 2) at 30 degrees, so the long points are opposite the notched edge, as with the bottom truss members. Do not bevel the ends of the 48" apex rafter. Position the rafters onto the trusses, roughly 16" apart. Drill pilot holes and fasten the rafters to the trusses with a 2" screw at each intersection. Install the 48" rafter by screwing through the outsides of the trusses and into the ends of the rafter. The 48" rafter is placed in order that the roofing material may be attached to the apex. The tops of all of the rafters should be ¾" above the tops of the trusses. Fill in the spaces between the rafters on the truss with 1×2s screwed to the trusses to create a flush surface for the corrugated roofing.

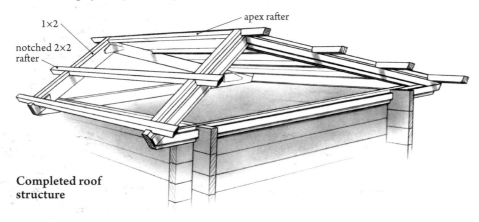

1×2

notched 2×2 rafter

apex rafter

Completed roof structure

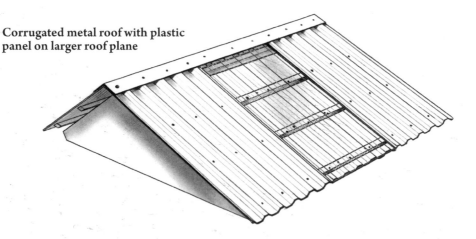

Corrugated metal roof with plastic panel on larger roof plane

10. INSTALL THE ROOFING.

On our coop, we used corrugated metal roofing, with a section of clear plastic roofing in the middle for extra sunlight in the coop. Cut and install the roofing so it overhangs the rafters by at least 1" on the sides and 2" at the ends of the roof structure. Fasten the panels with 1½" roofing screws (see page 26 for tips on installing corrugated roofing). Snap a chalk line across the roof centered over each rafter, and use this as a guide for your roofing screws. Do not screw the roofing panels to the apex rafter yet.

Cut, bend, and install a piece of 4" flashing to cover the apex of the roof (see page 37 for tips on bending sheet metal). To keep the flashing straight, we added sections of wiggle board between the corrugated metal and the flashing. Use the roofing screws to fasten through the flashing, wiggle board, and roofing panels and into the apex rafter.

11. HANG THE BARN DOOR.

There are a couple of options for installing the barn door. You can use prefab barn door tracks and hardware, which can be pricey. Or, you can do what we did and make your own with two custom-cut L-shaped 2×4s for tracks and cheap casters from the hardware store for door rollers.

Install the casters at the top and bottom of the door, using 2" screws. Cut an 8-foot 2×4 into two pieces at 47" each. Using a table saw and dado blade, rip each 2×4 to create an L piece with the dimensions shown in the *L-track detail*. Then, switch to a standard table saw blade and rip each piece to 2¼" in width, retaining the L part of the tracks. Once installed, these tracks should be wide enough to allow the door's casters to roll smoothly without pinching. If necessary, adjust the dimensions of the L-track to accommodate your casters.

Drill a few small holes through the bottom of one of the tracks to drain water; mount this lower track to the front wall with 3" screws, so the door will be centered vertically over the barn door opening.

Drop the door (Z side out) into the lower track, then mount the top track to capture the door. Leave a ¼" gap so the door rolls easily. It helps to have a buddy here to hold the door and track in place. Cut and install a 2×2 ledger underneath the bottom track to support the weight of the door.

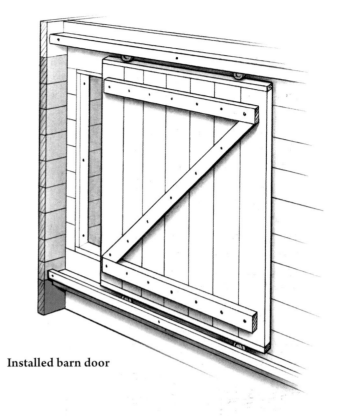

Installed barn door

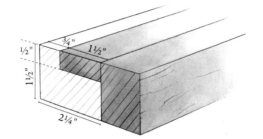

L-track detail

12. CREATE THE EGG ACCESS DOOR.

The left wall of the coop gets a small sliding door so you can easily reach in and gather eggs from the nesting box just inside the door. Cut the door opening at 7" × 10", plunge-cutting with a circular saw and finishing the cuts with a handsaw, as with the barn door opening. Cut and install a 1×2 on each side of this rough opening.

Cut the door to size at 12" wide × 9" tall, using leftover ¾" plywood. If desired, use a 1" Forstner bit to bore a shallow finger hole that makes it easier to open and close the door. Paint the door now, before installation, or finish it along with the rest of the coop in step 14.

Create a top and bottom door track from common 2×2s, cutting the pieces to length at 25" each. Using a router or a table saw and dado blade, mill a 1"-wide × ½"-deep groove along the full length of each track, centering the groove along one of the 2×2 faces. Fasten the bottom track below the door opening with 2" screws. Insert the door and install the top track, leaving a ¼" gap so the door slides easily but won't fall out. You can add paste wax to make the door slide smoothly.

13. ADD THE NESTING BOX AND ROOSTING BAR.

Cut two pieces of ¾" plywood into right triangles, each with two 16"-long sides. One of these is for the nesting box shelf; the other is an optional roof we added because chickens like dark, enclosed places for laying eggs. Mount the nesting

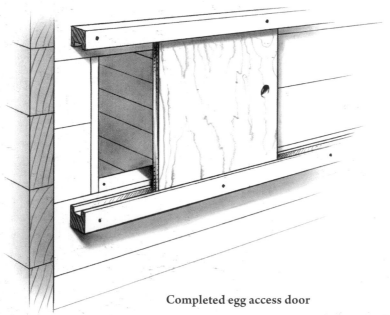

Completed egg access door

box shelf using 2×2 ledgers, so the top of the shelf is 1" below the opening of the egg access door opening (this prevents straw from bursting out every time you open the door). Add 1×2 edging to the front of the nesting box so eggs don't roll off. You can attach the nesting box roof from outside the coop with 2" wood screws (it helps to have a buddy here to hold it in place).

Cut a 2×2 roosting bar to length at 36", mitering the ends at 45 degrees. Install the bar about halfway between the coop floor and the nesting box shelf, using 2" screws. You can also add cleats below the roosting bar for support, as seen in the finished coop photos. The chickens will use the roosting bar as a step to hop into the nesting box.

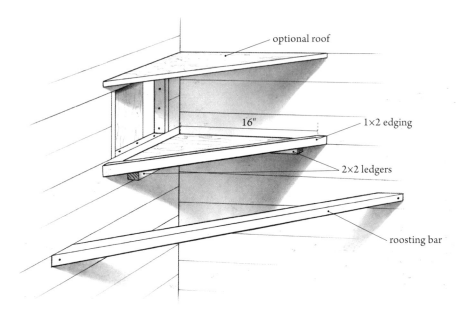

Nesting box and roosting bar

14. FINISH THE COOP.

Enclose the two open areas where the roof overhangs the side walls, using hardware cloth secured with 1×2s.

Install a handle and latch on the barn door and a latch on the egg access door (we used hasp latches for both doors; see Locks and Latches on page 27).

Finish the coop with two generous coats of natural deck sealant to all exterior surfaces of the cabin.

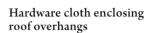

Hardware cloth enclosing roof overhangs

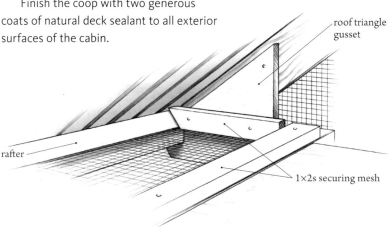

CHICK-IN-A-BOX

Chick-in-a-Box was our first coop as collaborators and holds a dear spot in our hearts. In 2010, we decided to make "something" for Maker Faire, a convention for DIY enthusiasts and tinkerers, to serve as an example of what our business is capable of and the scale at which we were hoping to build things. We also felt that we had something to *say* design-wise. A coop, we hoped, could be a conversation starter, prompting more in-depth discussions about food and self-sufficiency. It could also be a home accessory of immense beauty and integrity.

Matt had always wanted to build a structure with a butterfly roof, primarily because of the sense of wonder it inspires. "Why would you have your roof inverted like that?" and "Where does the water go?" were questions we wanted people to ask. The butterfly roof also harkens back to an era of Midcentury Modernism that we and many other Californians are particularly fond of.

So we built the coop, and it was a labor of love and experimentation. Along with the unconventional roof, we wanted to wrestle with compound angles and post-and-beam-style joinery. The frame is made up of redwood 2×4s and is notched and joined in such a way that it can come apart into four separate wall sections for easy transport from shop to site. The roof sits at an angle in two directions so that the water is diverted to the center on the low side, and there is a collection funnel to deliver the rainwater to the chickens.

Some other details we geeked out on include a cabinet-style nesting box with sliding doors, a "ladder" with half-lap joints for the chickens to climb, and a front door that folds down to convert to a ramp. Plywood and mesh were added to the exterior to enclose the volume, and Chick-in-a-Box was complete . . . just in time for Maker Faire.

We didn't know what to expect at Maker Faire. We were at home with others in the "sustainability" tent and had brought some examples of our furniture as well. A lot of people loved the coop, particularly kids who saw it as a giant playhouse, but nobody wanted to order one of their own. We realized that while building a handmade custom chicken coop was fun, there wasn't necessarily a market for this luxury item, so we didn't come away thinking we'd go into production. On the other hand, the coop lends itself to an industrious DIYer with modernist sensibilities, and more importantly, it was the catalyst for the rest of the designs in this book.

MATERIALS

- Eight 8-foot rot-resistant 2×4s
- Four 8-foot 2×4s
- Nine 8-foot 2×2s
- Seven 8-foot 1×2s
- One 5 × 5-foot sheet ¾" cabinet-grade plywood (Euro-ply; see page 25)
- Two 4 × 8-foot sheets ¾" marine plywood
- Fifty 2½" deck screws
- Fifty 1¼" deck screws
- Fifty 1½" roofing screws with neoprene washers

- 25 linear feet ½" galvanized hardware cloth, 16-gauge, 48" wide
- One 6"-wide × 4-feet-long metal V flashing
- Four 2 × 8-foot sheets corrugated metal roofing
- Five 1½" exterior hinges with screws
- One 1"-diameter copper pipe, 48" long
- One 90-degree copper elbow for 1" pipe
- Two 1" pipe straps, 6" long

- Exterior wood glue
- One funnel
- Metal bowl or catch pan (see step 12)
- 24" length ½"-diameter vinyl hose
- 1 gallon exterior paint
- 1 gallon clear exterior finish
- Three swivel gate latches with screws

SPECIALTY TOOLS
- 1" drill bit
- Clamps

1. CUT THE WALL FRAME PARTS.

Each of the four wall frames is made up of two 2×4 legs and two 2×4 crosspieces. The crosspieces fit into notches cut into the legs to form a ladderlike structure. The top of each leg piece has a compound cut that's easiest to make with a compound miter saw. When the frames are joined together, the legs at each corner will be doubled up; this means that you have to cut the mating legs together so that their tops are flush upon assembly. Due to the 3.6-degree bevel, the inside leg in each corner pair will be slightly shorter than the outside leg.

Cut the leg pairs for each corner, using rot-resistant 2×4s. Cut one pair to length at 63", with a 21-degree miter and a 3.6-degree bevel. Cut another pair to length at 63", with an 18-degree miter and 3.6-degree bevel. Orient the miter cuts so that the tops of the legs will slope toward

the center of the assembled coop, and the bevels so that the water flows from the high to the low side.

Cut the remaining two leg pairs to length at 66", with one pair at 21 degrees and one at 18 degrees, both with the 3.6-degree bevel toward the coop's center.

Cut the eight crosspieces to length at 45", using common 2×4s; all of these get square end cuts.

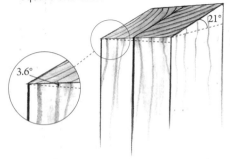

Doubled-up leg with compound cut

2. ASSEMBLE THE WALL FRAMES.

Notch the legs for the crosspieces, using a circular saw and chisel, a table saw, or a miter saw. The interior legs at each corner get 1½" × 3½" notches, while the exterior legs get 1¾" × 3½" notches; see *Frame sides A and B* and *Frame sides C and D*, at right.

Assemble each wall frame by setting the crosspieces into the leg notches and fastening the pieces together with 2½" screws, checking the frame for square as you go. Join the four assembled wall frames together to create the coop frame, using 2½" screws.

NOTE: *If you'd like to be able to disassemble the coop frame, locate the screws so they'll still be accessible after the siding and mesh are installed. Walls A and B will get plywood siding, which will cover some of your frame screws installed now; remove these screws when it's time to install the siding, then use them to secure the siding and wall frames.*

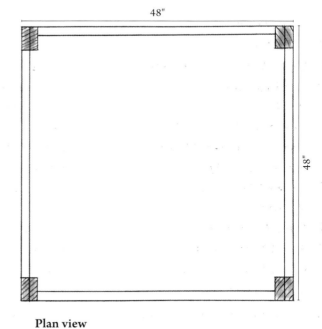

Plan view

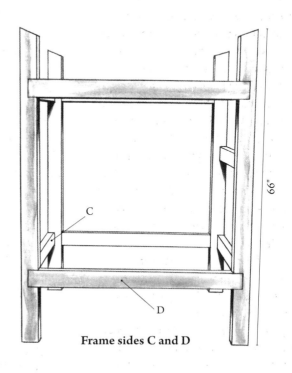

18°   21°

63"

51"

36"

A

B

12"

1¾"

45"

Frame sides A and B

66"

C

D

Frame sides C and D

3. CONSTRUCT THE ROOF FRAME.

The roof is made with 2×2s and consists of two sets of rafters and seven purlins. Cut and install the frame pieces following the purlin and rafter diagrams and the roof framing drawings here. Some of the angles on the rafters are too acute to cut on the miter saw, so cutting with a circular saw is recommended. Note the differences between sides B and D in relation to where the rafter pieces join each other to accommodate for the slope of the butterfly roof.

Each set of rafters should sit flush on the ends of the legs, but because they are not cut at compound angles there may be a small gap where the rafter meets the top crosspiece; a solid screw connection here should minimize such a gap.

Screw the rafters to the legs and crosspieces with 2½" screws. Then, starting 3" from the low point of the roof valley (where the rafters meet), install the purlins 12" on center, driving two 2½" screws at each intersection.

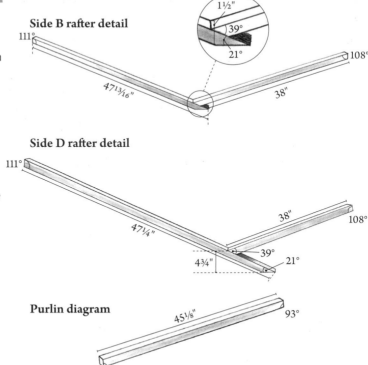

Side B rafter detail

Side D rafter detail

Purlin diagram

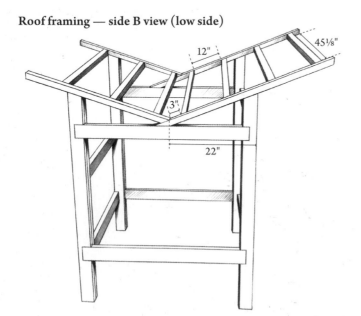

Roof framing — side B view (low side)

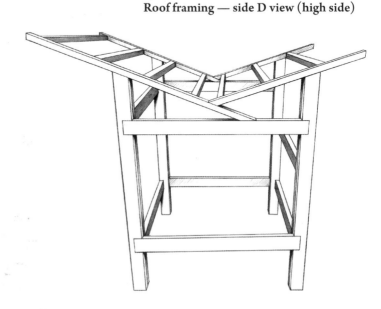

Roof framing — side D view (high side)

4. BUILD THE NESTING BOX CABINET.

Construct the nesting boxes out of ¾" Euro-ply plywood (see page 25). Cut the two sides, the bottom, and the two divider pieces as shown (*Nesting box interior view* and *exterior view*), using a table saw or a circular saw with a straightedge guide for accuracy. Also cut a 4"-wide × 42"-long piece for the top of the sliding door track.

Use a table saw or a router and ¾" straight bit to mill a ³⁄₈"-deep × ¾"-wide dado into the top track piece and into the bottom cabinet panel; these form the channels for the sliding doors.

Drill pilot holes and assemble the box cabinet as shown, using wood glue and 1¼" screws. It helps to clamp the corner joints to keep the parts from moving while fastening. Install the cabinet so it rests on the upper 2×4 crosspiece of frame side A, using 2½" screws. How far you want the cabinet cantilevered out of the coop is up to you. We chose a modest reveal of 3" on the coop exterior.

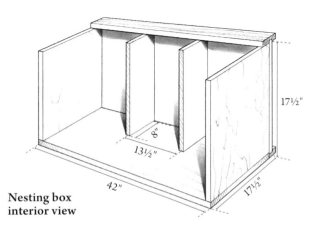

17½"
8"
13½"
42"
17½"

Nesting box interior view

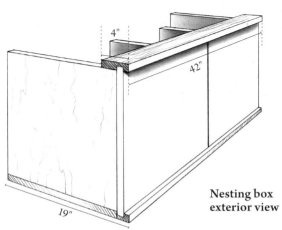

4"
42"
19"

Nesting box exterior view

5. ADD THE NESTING BOX DOORS.

Measure between the flats of the door channels in the bottom panel and top track piece of the nesting box cabinet, then subtract ¼"; this is your height for the two sliding doors. Cut a 48"-wide piece of the Euro-ply to this height, then rip the piece up the middle with a 30-degree bevel cut. The bevel ensures there's no gap when the doors are closed.

Install the doors by sliding them into the dado track. You can also add a little paste wax to the doors and tracks to keep the doors sliding smoothly.

6. INSTALL THE FLOORING AND SIDING.

Cut the floor from ¾" marine plywood, notching the corners to fit around the legs of the wall frames. Set the floor in place on top of the lower 2×4 crosspieces and attach with 1¼" screws, making sure its edges do not extend beyond the outside faces of the crosspieces.

Cut the siding panels to fit frame sides A and B, using the marine plywood. Side A gets two rectangular pieces, one above and one below the nesting box. Side B gets one piece that follows the shape of the roof. The bottom of the siding on both sides should be flush with the bottom edges of the lower 2×4 crosspieces. Fasten the siding to the frame members with 1¼" screws driven into countersunk pilot holes so that the screw heads are flush with the siding surface.

NOTE: *If you want the coop to disassemble easily into four sides, keep the edges of the siding well back from the outside edges of the corner posts. Also, if you plan on painting the plywood siding, it's easiest to do this before attaching it to the frame.*

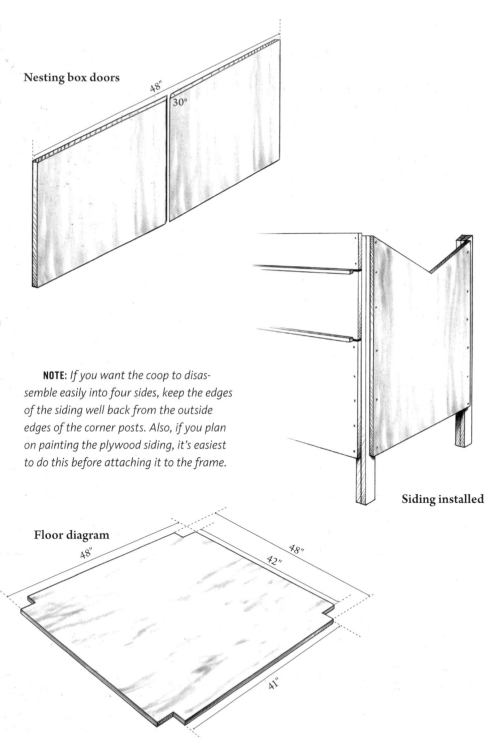

Nesting box doors

48"

30°

Siding installed

Floor diagram

48"

48"

42"

41"

41"

7. BUILD AND HANG THE FRONT SCREEN DOOR.

Side C of the coop has a large screen door and a smaller plywood door that doubles as an entry ramp for the birds. The doors are divided by a vertical 2×2 frame piece screwed between the upper and lower 2×4 crosspieces. Cut this 2×2 divider to fit, and install it so the larger door opening is 27" wide.

Build the frame for the screen door with 2×2s, making it ½" shorter and narrower than the rough opening in the wall frame. Butt the pieces together at the corners, and use two 2½" screws at each joint. Add hardware cloth to the outside of the door, securing it with 1×2s (see page 16). Hang the door with two hinges on the left side.

Screen door frame

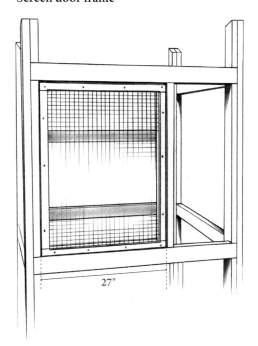

27"

NOTE: *On this coop, the doors are actually on the outside of the wall framing, but we think it is more secure to install them flush with the framing when closed.*

8. BUILD AND HANG THE RAMP/DOOR.

The ramp/door is made with ¾" marine plywood and consists of a flat plank wrapped with 1¾"-wide plywood strips, creating a ½" lip along the perimeter of the plank on all sides. The finished door should be ½" shorter and narrower than the rough opening in the front wall frame.

Cut the plank to size at 11½" × 33½" (or about 1" narrower and shorter than the rough opening). Miter the plywood strips to fit around the plank and fasten them with 1¼" screws so the strips are centered over the plank's edges. Hang the door off the lower crosspiece of the wall frame with two hinges at the bottom of the door.

Ramp/door with plywood trim

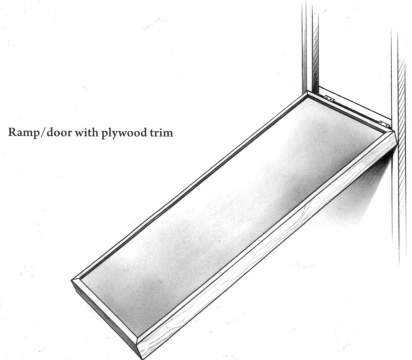

9. INSTALL THE HARDWARE CLOTH.

Enclose the top portion of side C and all of side D with hardware cloth, securing the edges with 1×2s and 1¼" screws.

10. COMPLETE THE ROOF.

Cut a single piece of galvanized metal V flashing to run along the valley of the roof and extend 2" beyond the edge of the rafters on the low side and flush with the edge of the rafters on the high side. Set the flashing in place but do not fasten it. Cut four pieces of corrugated roofing to length for each side so they overlap the flashing by at least 2" and overhang the top rafter ends by 2" (see page 26 for cutting and installation tips).

Starting at the high side of each roof plane, position the roofing panels so they overhang the rafters by 2" and overlap onto the flashing. As you add each successive sheet, slip the lower piece underneath the previous one to create a watertight joint. Cut the pieces at the low side of the roof to size.

Once all of the pieces are properly positioned, fasten the roofing with roofing screws with neoprene washers, driving the screws into the valleys of the corrugations where they make contact with the rafters and purlins.

11. BUILD THE LADDER.

The ladder allows the chickens to climb from the coop floor to the nesting boxes. It has 2×2 main supports and 1×2 rungs that are "let in" to the supports for a flush installation.

Cut the two 2×2 main supports to length at 33", mitering the bottom ends at 48 degrees. Cut six 1×2 rungs to length at 12". Using a table saw, a miter saw, or a circular saw and a chisel, notch the top faces of the main supports so the rungs fit snugly into the notches and are spaced 4" apart.

Assemble the ladder so all of the top and side surfaces are flush, and fasten the rungs to the main supports with 1¼" screws. Install the ladder by screwing through the main supports and into the floor and the bottom shelf of the nesting cabinet.

Cut and install a 2×2 roosting bar between the bottom of the nesting cabinet and the vertical post between the front doors.

Completed ladder

12. ADD THE WATER CATCHMENT SYSTEM.

The simple water catchment system collects runoff water from the lower end of the roof's valley flashing (at side D of the coop) and diverts it to a catch pan inside the coop.

Drill a hole through the plywood siding for 1" pipe to enter the coop 6" above the floor. Use a pipe strap to attach a funnel to the siding, directly below the end of the valley flashing. Run a length of straight 1" pipe from the funnel down to the hole, and add a 90-degree elbow to make the turn into the coop. Secure the pipe and elbow with pipe straps. There's no need to solder or otherwise secure the pipe joints.

Set a large metal bowl or catch pan underneath the elbow inside the coop to collect the water. Be sure to include an overflow hole at the top of the bowl to prevent an overflow from flooding the coop. Attach a 24" length of ½" hose to the overflow hole and run it down through the floor for drainage.

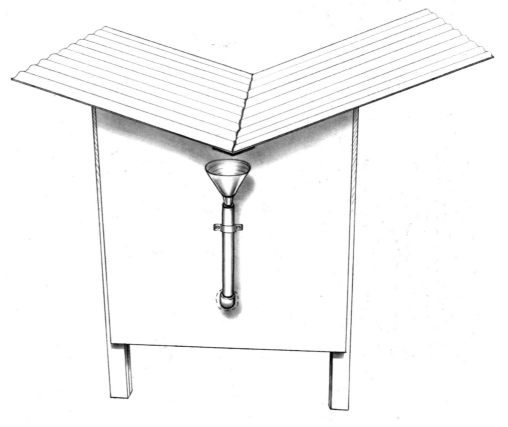

Water catchment system

13. FINISH THE COOP.

Finish all exterior wood surfaces of the coop with a clear coat or other protective finish. Install latches on all of the doors.

SYM

GUEST DESIGNERS: YVONNE MOUSER AND ADAM REINECK OF NEW FACTORY IN SAN FRANCISCO, CA

SYM is much more than a chicken coop; it's a symbiotic urban farming system. Its designers needed a coop that fit nicely into a small city yard. By combining multiple garden elements, they found that they could maximize the use of space while improving production and minimizing waste. The result is an elegant ecosystem of mutually beneficial components, integrating a chicken coop, greenhouse, worm compost bins, and roof-water collection.

The coop is large enough to happily house up to four chickens. It sits above two worm compost bins that are fertilized directly by the chickens' manure. Food and garden scraps can go either to the chickens or into the worm bins to create nutrient-rich compost for the garden. The chickens can eat insects that are naturally attracted to the compost and come in through the coop's mesh floor, as well as worms taken from the bins and fed to them, adding essential proteins to the hens' diet to encourage the production of rich and delicious eggs. The rooftop collects rainwater that can be directed to the chickens for drinking or to a storage container for future use.

The frame-and-panel construction of the SYM coop is part of a modular design concept that allows for adding to, or adapting, structures and other elements to accommodate greater production. Some additions include a chicken run, a raised greenhouse, additional water storage, and a cold frame (see http://newfactorysf.com/sym). Each element enhances the natural symbiotic relationships within the system to eliminate waste and improve production.

SYM's designers operate New Factory in San Francisco, California, a design/build studio focused on creating and reinterpreting everyday objects in order to engage people in new behaviors and meaningful interactions. SYM was originally created for a backyard gardening competition and was built over the course of a couple of weekends, using off-the-shelf parts from a local hardware store. Painting the galvanized metal frame is not necessary, but the designers wanted to give the coop a little color. If you can't find galvanized steel, it's best to paint the steel with two or three coats of exterior paint. The SYM coop resides in the designers' backyard in San Francisco, among the raised garden beds and woodstove.

MATERIALS

- Four 6-foot pieces 1¼" × 1¼" galvanized perforated steel angle
- Eight 4-foot pieces 1¼" × 1¼" galvanized perforated steel angle
- Three 4 × 8-foot sheets ½" marine grade plywood
- One 8-foot rot-resistant 2×4
- Three 8-foot rot-resistant 1×6s
- One 12-foot rot-resistant 2×4
- One 12-foot 2×4
- One 10-foot 2×4
- Two cans spray paint primer
- Two cans spray paint in yellow (or color of your choice)
- One gallon exterior primer
- One gallon exterior paint in white (or color of your choice)
- Sixteen ¼" × ¾" #20 hex cap bolts
- Seventy ¼" #20 nuts
- Seventy ¼" flat washers

- Fifty ¼" × 1¼" #20 hex cap bolts
- Twelve 1½" galvanized pan-head wood screws with galvanized washers
- Sixty 1½" deck screws
- Twenty-five ¾" deck screws
- One box 1¼" exterior finish nails
- Ten 2½" deck screws
- Fifty 1½" #9 hex-head roofing screws with neoprene washers
- Twelve linear feet ½" galvanized hardware cloth, 18-gauge, 48" wide
- One box ⅜" galvanized staples
- Ten 2½" exterior butt/mortise door hinges with screws
- Six 2½" exterior barrel bolt latches with screws
- One 48" piece gutter with end caps and drain spout
- One 10-foot piece 2×2 galvanized drip-guard flashing

- One 10-foot wiggle board
- One 8-foot wiggle board
- One 10-foot and one 8-foot piece corrugated metal roofing, 26" wide
- 2" galvanized pipe and fittings (as needed; see step 19)
- Two 50-gallon plastic storage bins (2 × 4 × 2 feet) (Note: Buy the sturdiest ones you can find so they don't deform with the weight of the compost.)
- One 5- to 20-gallon water tank (or larger, as desired)

SPECIALTY TOOLS

- #9 hex driver bit
- Staple gun (preferably round crown-type)
- Two ⁷⁄₁₆" wrenches (or adjustable wrenches)

1. PAINT THE STEEL ANGLE.

This step is optional, provided the steel angle is made of corrosion-resistant material. The designers used a nice egg-yellow color found at their local hardware store. Paint all 12 of the steel angle pieces, using two coats of spray primer and two coats of spray paint. Follow the manufacturers' directions for application and drying times. Let the paint dry thoroughly.

2. CUT THE PLYWOOD PANELS.

Cut three 4 × 8-foot sheets of plywood in half to make six 4 × 4-foot panels, using a circular saw with a straightedge guide to ensure clean, straight cuts. Four of these panels will become the front, back, right side, and left side walls of the coop (with door and window cutouts made in the next step). Cut two of the 4 × 4-foot pieces in half to make four 2 × 4-foot panels; three of these will become the back, right side, and left side of the bin enclosure below the coop. Reserve the remaining 2 × 4-foot piece for the nesting shelf (step 12).

NOTE: *If the full plywood sheets measure exactly 48" × 96", make your cuts directly down the centers of each panel, so that the cut pieces are the same size. Most circular saw blades cut a kerf (the material removed by the blade) of slightly less than ⅛", meaning that the resulting pieces will be short by ¹⁄₁₆" or less — a negligible amount.*

3. MAKE THE DOOR AND WINDOW CUTOUTS.

Lay out the door and window cutouts on the four coop wall panels, as shown in the *Coop wall cutting layouts*. Note that all of the wall panels, windows, and doors have 2½" borders. The material removed for the door openings will be used to create the doors, so make these cuts carefully; support the cutout pieces so they won't fall away and become damaged or pull off large splinters as you complete the final cuts at the corners. The door in the back wall is a solid panel with no window cutout.

Make the cutouts with a circular saw, using the plunge-cutting technique described on page 34. Finish the cuts at the corners with a handsaw or jigsaw. Handsaw and jigsaw kerfs typically are smaller than those of a circular saw blade, so the edges of the cutouts and/or the cutout panels may need to be sanded flush at the inside and/or outside corners. Sand down the excess material using coarse sandpaper and a sanding block to maintain straight edges. Save all scrap material for the nesting boxes and other details.

Coop wall cutting layouts

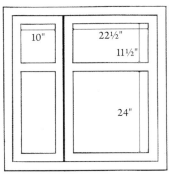

Front

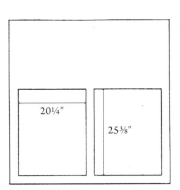

Right side

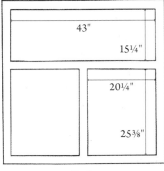

Back

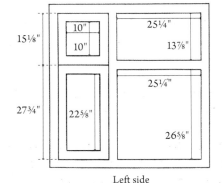

Left side

4. PAINT THE PLYWOOD PIECES.

Sand all surfaces of the coop walls, doors, and bin enclosure panels. Apply two coats of exterior primer to all of the pieces, following the manufacturer's directions. Apply two topcoats of exterior paint in the color of your choice. You can be as creative as you'd like with the color on your coop. SYM's designers chose white to create a nice contrast to the natural wood elements and bright-yellow angle framing.

5. ASSEMBLE THE COOP FRAME.

Lay down two of the 6-foot steel angle pieces so they are parallel to each other; these are the vertical supports for one side of the coop frame. Position two 4-foot angle pieces across the vertical supports so one is at the top ends of the vertical supports and one is 24" from the bottom ends of the verticals; these are the horizontal supports.

NOTE: *Be sure to orient all of the 6-foot vertical angles in the same way: The measurements from the last hole to the end of the angle may differ between the top and bottom ends of each piece. Either orientation is fine; just be sure they are consistent.*

Bolt the angle pieces together with four ¼" × ¾" #20 hex cap bolts, washers, and nuts. Make sure the horizontal angles have one flat side facing up, as shown in the *Frame bolt detail*, because these will eventually support the floating floor and roof structure above. Repeat the same process to assemble the other side frame for the coop.

Join the two side frames with the four remaining 4-foot angles, bolting the joints as before. As shown in the detail, the horizontal angles overlap each other at the inside corners of the frame. Arrange the overlaps so the horizontal supports at opposite sides of the frame are the same, ensuring that they are level and at the same height with each other.

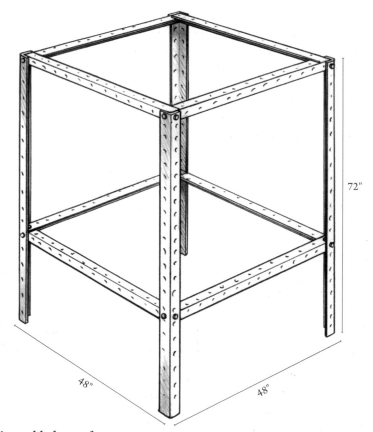

72"

48"

48"

Assembled coop frame

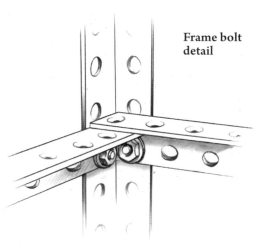

Frame bolt detail

6. INSTALL THE BIN ENCLOSURE PANELS.

Position the three 2 × 4-foot painted plywood panels on the inside of the coop frame, below the lower horizontal supports. Leave one side open to become the front of the coop. Mark the top corners of the panels, and notch or clip the corners as needed to provide clearance for the frame bolts, as shown. Prime and paint the cut edges of the panels.

Reposition the enclosure panels (starting with the back panel, then butting the side panels against the back), and drill three ¼" holes evenly spaced along the top of each panel, drilling through the holes in the lower horizontal supports. Then, drill one hole through the vertical support, 6" up from the bottoms of the panels on each side edge. Install each enclosure panel with five ¼" × 1¼" #20 hex cap bolts, washers, and nuts.

NOTE: *To prevent tearout on the interior plywood face (when drilling from the outside), clamp a block of scrap wood to the inside face, over the hole location. Alternatively, you can predrill the hole from the inside, using a smaller bit, then drill the final hole with the ¼" bit, going from the outside in.*

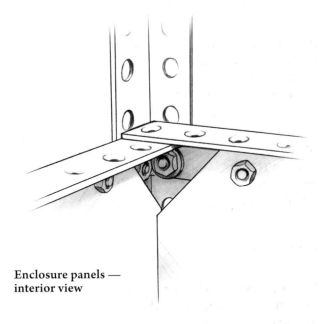

Enclosure panels — interior view

7. ADD THE BASE RUNNERS.

Cut two runners to length at 47½", using an 8-foot rot-resistant 2×4. Position the runners against the inside faces of the side bin enclosure panels so the runners' bottom edges extend ½" below the bottom ends of the coop's vertical frame supports. Fasten each runner by screwing from the outside of the side enclosure panel with six 1½" galvanized pan-head wood screws and washers.

Base runners installed

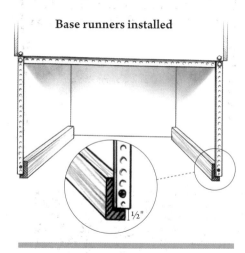

Mesh cutting diagrams

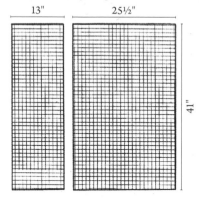

Front

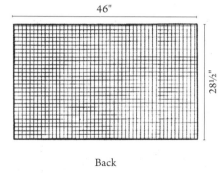

Back

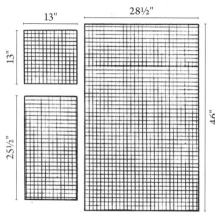

Right side

Left side

8. ATTACH HARDWARE CLOTH TO THE WINDOWS AND DOORS.

Cut seven pieces of hardware cloth mesh to cover the window openings in the coop's walls and doors, as shown in the *Mesh cutting diagrams*. The mesh will overlap onto the plywood edges by about 1½" on all sides. Attach the mesh to the inside faces of the wall and door pieces, using staples driven about every 5". Any heavy-duty staple gun will work, but a round crown-type gun is ideal because the tip registers on the round wires.

9. INSTALL THE DOORS AND LATCHES.

Lay the four wall panels on a flat surface with their outsides facing up (mesh facing down). Install the doors with 2½" hinges and bolt latches as shown. Take care to ensure an even gap around all of the doors' edges. Use a paint stick or a couple of quarters to shim the doors and ensure consistency before screwing in the hardware. The spacing will be determined by the width of your circular saw blade (the kerfs from the cutouts made in step 3). Hold off on installing the narrow (left) door panel on the front wall until after step 13.

Door hinge and latch locations

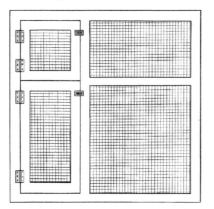

Left side

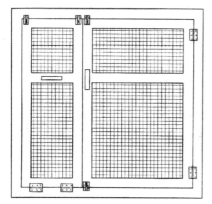

Front

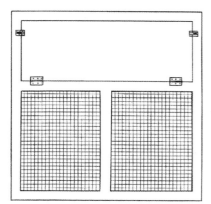

Back

10. ATTACH THE WALL PANELS TO THE COOP FRAME.

With the help of a friend, hold or clamp each assembled wall panel to the *outside* of the coop's steel frame. Drill eight ¼" holes into each wall panel, drilling from the inside through the holes in the steel angle. Drill one hole at each corner and one hole centered at 24" in between.

NOTE: *Remember, you can prevent tearout in the plywood by predrilling with a smaller bit from the inside out, then using the ¼" bit from the outside in or, if possible, by clamping a block at the back side of the hole location.*

Bolt each wall assembly to the frame with eight ¼" × 1¼" #20 hex cap bolts, washers, and nuts. Check all of the doors to make sure they open and close properly, and make any adjustments as needed.

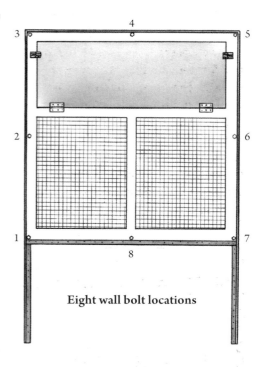

Eight wall bolt locations

11. BUILD THE FLOOR.

Cut three pieces of rot-resistant 1×6 to length at 48", and cut six pieces to length at 15". You should have one 48" piece left over; save this for the nesting shelf (step 12). Cut three pieces of rot-resistant 2×4 to length at 44".

Arrange all of the pieces on a flat surface as shown. Be sure to leave 2" of space between the 2×4s and outside edges of the 1×6s. Fasten the assembly with 1½" deck screws driven through each of the 1×6s and into the 2×4s below. Use four screws in each 15" piece and two screws where the 2×4 crosses the 48" 1×6s.

Cut off 1" of each corner of the frame at a 45-degree angle to provide some wiggle room for placing and removing the floor inside the coop. Cut a 48" × 48" piece of hardware cloth and staple it securely to the top side of the floor frame.

NOTE: *If you don't plan to let your hens out often to scratch in the dirt, you can cut out one of the squares of the floor and add a second ladder (page 115) to create a three-story coop. Cover the opening of the base of the coop with mesh so you can watch them scratch in the dirt below.*

12. BUILD THE NESTING SHELF AND ROOSTING BAR.

Using the leftover 2 × 4-foot plywood panel, cut a 1½" strip off one long side of the panel. Cut a 1½" × 21" strip from another leftover piece of plywood. The panel will become the nesting shelf, and the two 1½" strips will serve as cleats for supporting the side and rear of the shelf. Notch 1½" off the two back corners of the shelf panel at 45 degrees.

Rip the leftover 48" 1×6 (from building the floor) down the middle lengthwise to create two equal pieces about 2¾" wide, using a table saw or circular saw. These will be used as a ledger for the front edge of the nesting platform, and for the roost.

Install the 1½" × 48" plywood cleat along the back of the coop wall, just above the lower windows, using six ¾" deck screws on each panel. Install the 1½" × 21" cleat along the right wall, just above the lower windows and so that its top edge is level with the rear cleat's. Similarly install the left-side plywood cleat.

With a helper, hold the 1½" × 48" cleat to the front of the 21" cleats so it spans between the left and right walls of the coop, with its bottom edge flush with the bottom edge of the cleat. Fasten the 1×3 from the outside of the coop with two 1½" deck screws on each side so that the top is flush with the top of the cleats. Set the nesting shelf panel onto the cleats and drill through the panel into the cleats to secure it. As an option, you can attach a piece of the ripped 1×3 to the front cleat to create a lip above the nesting shelf; this will prevent the nesting boxes from sliding off the shelf. Fasten the back corners of the nesting shelf into the plywood cleats below, using 1½" deck screws.

Use the remaining 1×3 for the roost. Have a helper hold it level and parallel to the nesting shelf, about 12" toward the front of the coop, and attach it from the outside with two 1½" deck screws on each end.

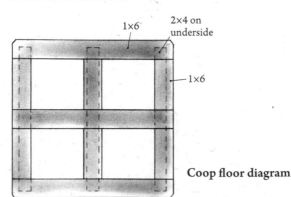

Coop floor diagram

Nesting shelf and roosting bar

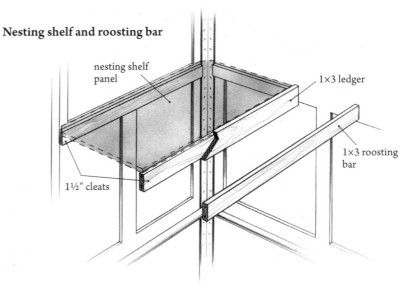

13. CREATE THE LADDERS.

Cut an 8-foot rot-resistant 1×6 in half to make two 48" pieces. Using a table saw or circular saw, rip four 1" × 48" strips from one of the 48" pieces. Cut the strips into 13 pieces at 5½" each and 7 pieces at 13½" each to make the ladder rungs.

Arrange 12 of the 5½" rungs evenly across the remaining 48" 1×6, and fasten the rungs with 1¼"-inch exterior finish nails. This completes the ladder that runs from the coop floor to the nesting shelf. Set the ladder at the desired angle, then fasten the last 5½" rung to the coop floor to serve as a stop that holds the ladder in place.

Arrange the 13½" rungs evenly along the inside of the narrow (left) door panel on the front wall, and fasten them with finish nails (the rungs aren't shown in the photos here; the designers added them later to give the chickens sure footing). Install the door to the front wall of the coop, using 2½" hinges at the bottom of the door. When the door is swung down it serves as a ladder for the birds to go in and out of the coop.

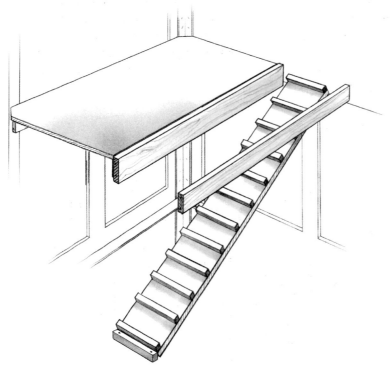

Completed nesting shelf ladder

14. BUILD THE NESTING BOXES.

Use the leftover plywood and rot-resistant lumber to construct a set of nesting boxes for the nesting shelf. See page 17 for help with determining the size and configuration of your nesting boxes. This coop has three 12"-wide × 9"-deep × 15"-tall boxes built as a single unit. To contain the bedding, attach a 48"-long piece of leftover wood on the shelf at the door opening.

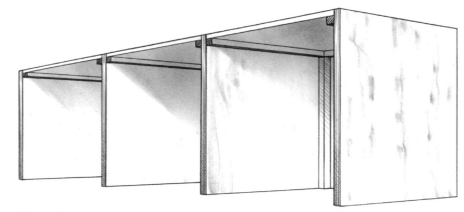

Nesting boxes

15. ADD THE GUTTER.

Prepare the 4-foot length of gutter by attaching a cap at each end and installing the drain spout (if necessary), following the manufacturer's directions. Remove the top two corner bolts from the right side of the coop frame. Clamp or hold the gutter in place on the right side of the coop so that the top of the gutter is just below the top edge of the coop frame. Drill through the existing two ¼" open holes from the inside of the coop out and through the gutter at each top corner. Attach the gutter to the coop frame, using the two 1½" existing corner bolts with a washer and nut on the inside of the coop.

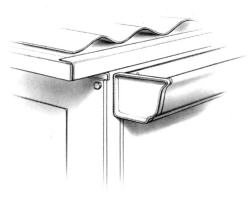

Gutter installed
(roof shown for clarity)

16. CONSTRUCT THE ROOF FRAME.

Cut two pieces at 60" long from a 10-foot 2×4. Cut three pieces at 45" long from a 12-foot 2×4. On two of the 45" pieces, use a straightedge and pencil to draw an angled line to taper the board's width from 3½" (full width) at one end to 2" at the opposite end. Cut along the line with a circular saw. If you live in a snowy climate, you should adjust the design to create a steeper roof slope.

Rip one of the 60" pieces to make it 2" wide. Assemble the roof frame as shown, using two 2½" deck screws at each joint.

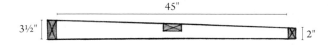

Roof frame diagram — front elevation

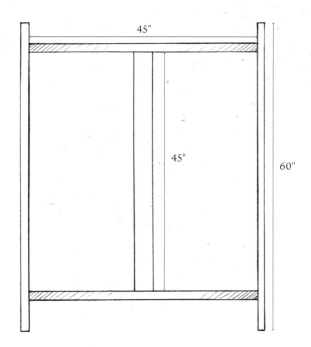

Roof frame diagram — top view

17. PREPARE THE ROOF FLASHING.

Cut four pieces of 2×2 drip-guard flashing to length at 60". Cut the upper flange at each end of two of these pieces, and bend the ends at 90 degrees so the resulting pieces fit over the 48"-long sides of the roof frame.

Shaping flashing for short roof sides

18. COMPLETE THE ROOF.

Cut three pieces of wiggle board to length at 60". Be sure that the waves in the wiggle board align with one another so that they will register with the waves in the corrugated roofing. Fit the flashing onto the roof frame as shown, then add the wiggle board pieces, centering them over the crosspieces of the roof frame. Secure the assembly by driving 1½" deck screws through the valleys (low points) of the wiggle board and into the roof frame

pieces, using four screws for each wiggle board.

Cut three pieces of metal roofing to length at 54". Center one piece of roofing on top of the roof assembly so its lower end (at the gutter) overhangs the lower drip-guard flashing by 1". Position the other two roofing pieces at either side of the center piece, overlapping the center piece so the outside edges of the outer roofing pieces overhang the roof structure

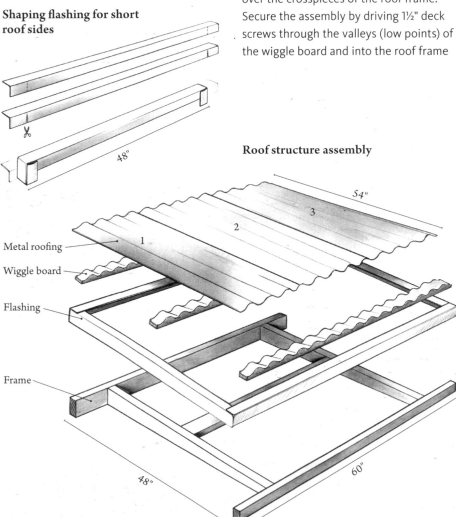

Roof structure assembly

Metal roofing

Wiggle board

Flashing

Frame

48"

54"

60"

48"

by about 4". The roofing will overhang the upper edge of the roof structure by about 4".

Secure the roofing with eight 1½" roofing screws on both the right and left sides of the roof, and six screws along the center, driving the screws in the valleys of the corrugation, through the wiggle board and into the roof frame pieces.

19. HOOK UP THE RAINWATER CATCHMENT SYSTEM.

Assemble 2" drain pipe and fittings to connect the gutter drain spout to the water storage tank. You can use any suitable tank for storing the water, but preferably look for something that holds more than 20 gallons and has two standard plumbing fittings for attaching a gutter pipe and a garden-hose spigot. SYM's designers built a custom stand for their water tank, using some of the leftover plywood from the wall panels. If you make a stand, be sure it's tall enough to allow for filling buckets or watering cans.

Place two 2 × 4-foot plastic storage bins into the bin enclosure space under the coop floor, and get ready for your chickens!

CORNER COOP

The Corner Coop was an experiment in how to incorporate the existing elements of a home or landscape into the coop structure. In this case, the corner of a tall privacy fence provides two prebuilt walls (including posts with concrete footings) to tie into. There was also a tangential experiment involving a cantilevered "floating roost," which is designed for both visual interest and easy cleaning.

The story behind the coop design is that the homeowner, Leah, came to us with a request for an "all-inclusive" coop for her hens. It should provide the chickens with plenty of space within a totally enclosed coop, so they wouldn't need to be let out in the morning or put away in the evening. As it turned out, designing the coop was a tricky affair. We didn't want the roof to go above the fence line, but Leah also wanted to be able to walk inside without bending over. The solution was to slope the roof very gradually, with the lowest point at 6 feet and the highest at 6 feet, 8 inches. You can get away with this in California but need a steeper incline in snowy climates.

Generous light was also a guiding factor. We used clear corrugated plastic roofing, which acts like a giant skylight, and used plenty of mesh on the exterior. The lower portion of the two framed walls is entirely meshed, allowing the chickens to look out and humans to look in.

Taking something that is usually horizontal or vertical and tweaking it with irregular angles is a good way to inspire debate and frustration. It makes the project increasingly complex and forces you to do math you haven't done since the seventh grade. At the same time, though, it can make the finished product so much more interesting, particularly when there is a good reason for doing so. In this coop, the front corner is where the eye is drawn when entering the backyard from the house. We wanted this corner to take a central role and be the most revealing about what was inside. The mesh comes to its low point at this corner, and the cedar takes over from there, going back toward the fence. The end result is a highly breathable, airy but ultra-secure coop.

And then there's the "floating roost." While Leah prefers to use the deep bedding technique (see page 13) with the coop floor, the roosting space inevitably needs to be cleaned out. So we wondered, "What if the roosting space were on a slope coming out from the coop so you could easily sweep it into a compost bin? And what if such a roost, which must also be very safe, were a geometric aberration from the rest of the coop, and from time and space itself?" Okay, so it's not a black hole, but it is intended to bring up some feelings of wonder. The cantilevered design also makes the roost work like a chute for quick and easy cleanup.

A note about the coop construction

Our goal with the following steps is to outline the basic techniques used to build the original Corner Coop. However, due to numerous site details and other design factors specific to each builder's plan, your coop might be very different from ours. The coop shown here is relatively easy to build. Though we suggest that the difficulty level is intermediate, it could become advanced depending on the design work required to accommodate each site. We also assume that if you're comfortable modifying the design to suit your needs you don't need as many nitty-gritty details of construction, so the steps are written for those with knowledge of basic carpentry techniques.

MATERIALS

> Twenty-three 8-foot 2×4s
> One 8-foot rot-resistant 2×4
> One 16-foot rot-resistant 2×4
> Two 16-foot 2×4s
> Eight 8-foot 2×2s
> Four 8-foot 1×2s
> Three 8-foot rot-resistant 1×2s
> Thirty 8-foot 1×6 cedar siding boards (or other siding material)
> Ten ⅜" × 3½" lag bolts with washers

> Fifteen 2×4 joist hangers with fasteners
> Two hundred fifty 1¼" deck screws
> One hundred fifty 2½" deck screws (for general framing)
> Thirty-two 1½" roofing screws with neoprene washers
> Four 8" × 16" × 2"-thick (or 4"-thick) solid concrete blocks (see step 3)
> Three 2½" exterior hinges with screws
> Two 1½" exterior hinges with screws

> Four 2 × 8-foot sheets clear corrugated roofing
> Four square feet opaque roofing material (see step 5)
> 40 linear feet ½" galvanized hardware cloth, 18-gauge, 36" wide
> Two exterior door latches with screws

SPECIALTY TOOLS

> Miter saw
> Ratchet set

1. ASSESS THE FENCE.

Thoroughly examine the existing corner of fencing where you intend to build the coop. A suitable structure should be at least 6 feet tall and have posts anchored in concrete no more than 8 feet apart. Inspect the posts for rot or other signs of potential weakness, and make sure all stringers and other framing members are structurally sound and securely fastened to the posts. Reinforce framing connections as needed with additional screws, lumber, or metal framing connectors. If you share any of the fence sections with a neighbor, make sure your plans are *coopacetic* with everyone involved before you start building.

2. INSTALL THE ROOF LEDGERS.

Determine the height and slope of the roof. On our coop, we chose to have the lowest point at 6 feet so that someone could easily walk into the coop and stand up throughout. The slope is a gentle 2 degrees, but again, we recommend altering this to take into account snow loads if you live in such a climate.

Cut and install 2×4 roof ledgers at the desired finished height of the roof, minus the height of the roofing material. Anchor the two ledgers to the fence posts with lag bolts. On our coop, the roof (and thus the ledger) slopes over the long dimension (front wall) of the building, while the ledger over the end wall at the fence corner is level. Run your ledgers as long as needed to reach the nearest fence post for anchoring. If desired, you can install additional ledgers for anchoring the bottom ends of the walls, depending on your situation.

Ledgers installed on fence posts

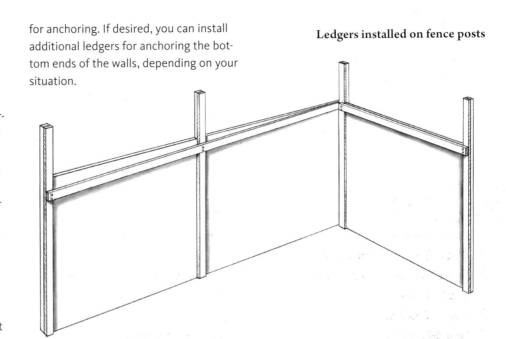

3. FRAME THE WALLS AND ROOF.

The two framed walls stand on concrete blocks for stability and to forestall rot. On our coop, the tops of the blocks are about 2" above grade; you can adjust this as desired. Mark the footprint of the coop onto the ground, then set down a solid concrete block at the outside wall corner and every 40 inches along the front wall. Use a level and a long, straight board or a mason's line to make sure the blocks are level with one another.

Frame the two walls with common 2×4s, but use rot-resistant lumber for the bottom plates. Space the studs 20" on center, or 16" on center if you plan to add insulation. Cut the top of each stud in the front wall to match the roof slope. Plan the first full stud space (nearest the fence-side end wall) to accommodate your roost frame. Our roost fits into a standard stud space, but you can modify yours as desired. Frame the coop door opening to the desired size; using doubled studs at each side if the opening is wider than 20".

Note that all of the framing (except for the bottom plate) on the end wall is the same orientation as the studs in the front wall.

Cut the 2×4 rafters to span between the long ledger and front wall plate. Install the rafters with joist hangers and the manufacturer's specified fasteners, aligning the rafters with the front-wall studs.

Construct a frame for the coop door, using 2×4s and making it ⅛" smaller than your rough opening on all sides. Hang the completed door frame with three 2½" hinges (see the section on hanging doors, page 30).

Install horizontal 2×4 blocking in all of the stud spaces, creating a 21" opening for the meshed windows. Raise and join the two wall frames, then anchor the top plates to the rafters.

4. FRAME THE ROOST.

The roost starts as a simple rectangular frame made with 2×2s. Size the frame to fit snugly into the first full stud space near the outer end wall. To create the floating effect, mount the roost frame to the front-wall studs, then add vertical 2×4 blocking sandwiched below and above the 2×2 framing. Install a couple of 2×2 knee braces to bolster the interior end of the roost, and any additional framing as needed until the installation feels solid. We made our roost pitch downward slightly at the front so it's easy to sweep out the contents into a compost bin.

NOTE: *Depending on your site configuration, you may want to install siding on the fence side of the roost before mounting the roost into the coop wall (see step 6).*

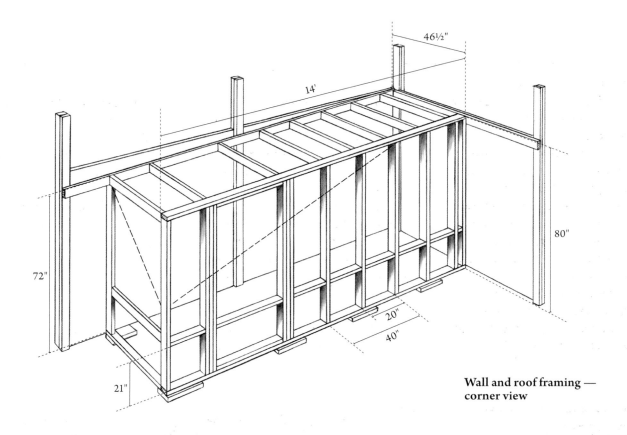

46½"

14'

80"

72"

20"

40"

21"

**Wall and roof framing —
corner view**

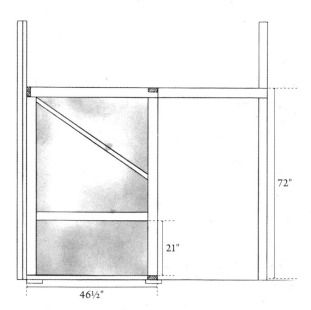

72"

21"

46½"

Wall and roof framing — end view

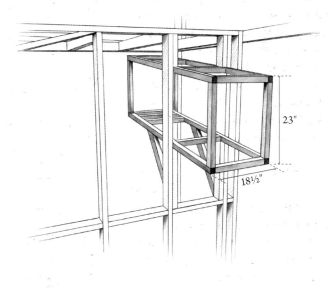

23"

18½"

Roost framing diagram

5. INSTALL THE ROOFING.

Cut and install the clear roofing panels on the coop so the corrugation runs parallel to the front wall (see page 26 for tips on working with corrugated roofing). Slip the lower pieces underneath the upper ones for proper drainage. Fasten the panels to the coop framing with roofing screws with neoprene washers.

Cover the exterior portion of the roost with opaque roofing so that the space feels more secure to the birds.

6. ADD THE SIDING AND MESH.

The siding and mesh window details give the coop its distinctive look, so it's worth the trouble of angling the top edges of the siding. For a special touch, use the same piece of siding to wrap around the front corner with a miter joint, creating the effect of continuous grain.

Snap chalk lines to represent the top edge of the siding on the front and framed end walls. Cut and install 2×2 nailers between the studs, with their bottom faces on the chalk lines; these will support the hardware cloth and provide a fastening surface for the 1×2 trim. Depending on your siding material, you may need to add extra blocking between the studs for additional backing and nailing surfaces.

Install the siding horizontally with 1¼" deck screws, trimming the upper pieces so they're flush with the nailers. We ran the siding over the door, then snapped chalk lines where the opening should be and plunge-cut with a circular saw. Before plunge cutting, make sure there is adequate framing for your siding so it will be supported during and after your cuts. Also enclose the exterior portion of the roost with siding.

Cover the remaining openings with hardware cloth, extending the mesh 12" below the walls and burying it in the soil in plane with the walls (see pages 16 and 26). Secure the mesh with 1×2 trim and 1¼" screws.

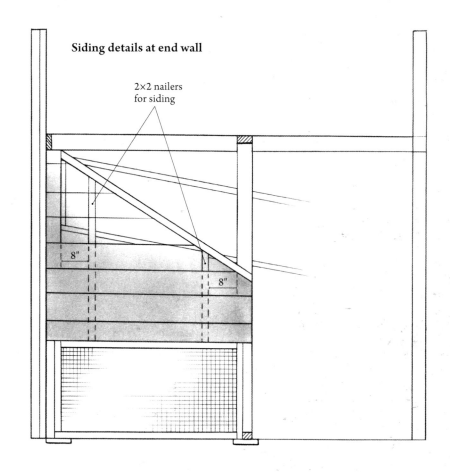

Siding details at end wall

2×2 nailers
for siding

8"

8"

---

7. ADD THE FINISHING DETAILS.

Create a floor for the roost with leftover siding planks. Build a door for the roost, using pieces of siding and 1×2 cleats or metal brackets to join the slats. Install the door with two 1½" hinges.

Build a ramp for the coop interior that starts near the door and runs up to the roost, using 2×4 supports screwed into the coop framing and a couple of pieces of siding for a plank, topped with 1×2 rungs spaced about 6" apart. Install latches (and locks, if desired) to secure the coop and roost doors.

Sand and finish the coop as desired. A good sanding cleans everything up, and an oil finish on the cedar will give it that extra pop.

KIPPEN HOUSE GARDEN ROOF CHICKEN COOP

GUEST DESIGNER: TRACI FONTYN OF KIPPEN HOUSE IN SEATTLE, WA
©2010 TRACI FONTYN

The designer of the Kippen House wanted to have a garden *and* chickens but lacked the yard space for both. So she had the inspired idea of putting the two together by giving the coop a garden bed for a roof. Another clever feature of this design is its modular plan, which makes it easy to build and add to in the future. Many of the pieces are the same size, and you can cut and paint almost everything prior to assembling the coop. The basic structure of the coop consists of six wall panels, which start out as identical lumber frames. Each frame gets trimmed out to suit its specific purpose, then the panels are joined together to create the coop's finished walls.

Location and site preparation are especially important for this coop. It should sit on solid, level ground and be positioned for plentiful sun exposure for the plants in the garden roof. The roof, by the way, has an added benefit beyond providing planting space: It's also an insulator that helps keep the coop interior cooler in summer and warmer in winter.

The Kippen House is perfect for up to three chickens. Its modern aesthetic and compact design are tailored for urban farmers who lack yard space and want a chicken coop that looks right at home in the city.

MATERIALS

- Two 10-foot rot-resistant 2×4s
- Seven 10-foot 2×4s
- Three 8-foot 2×2s
- Three 8-foot 1×12s
- Seven 8-foot 1×4s
- Three 8-foot 1×2s
- Twenty-one ⁹⁄₁₆" × 4" × 6-foot #2 cedar "no hole" square-top pickets
- One 4 × 8-foot sheet ½" plywood
- One 12"-diameter × 12"-long concrete tube form (Sonotube or similar)
- 10 linear feet ½" galvanized hardware cloth, 18-gauge, 36" wide
- One gallon exterior primer
- One gallon exterior paint (yellow, see step 1)
- One quart exterior paint (red, see step 1)
- Seventy-five 3" deck screws
- Seventy ¾" #8 pan-head screws
- Wood glue
- Two hundred fifty ¾" staples (for pneumatic staple gun; see Staple with Ease at right)
- Two hundred fifty 1¼" staples (for pneumatic staple gun)
- Six #10 exterior screw eyelets
- Seventy-five 1¼" deck screws
- Thirty 2" deck screws
- Two #10 screw hooks

- One 2" exterior barrel-bolt latch with screws
- Four ⁵⁄₁₆" × 3½" galvanized hex bolts
- Sixteen ⁵⁄₁₆" × 3" galvanized carriage bolts
- Twenty ⁵⁄₁₆" galvanized washers
- Twenty ⁵⁄₁₆" galvanized nuts
- Two 2" exterior hook-and-eye latches
- Eight Simpson Strong-Tie RTA12 corner angles
- Six 3" galvanized hinges with screws
- Four 3" × ⅝" Everbilt 3-surface corner braces (see step 7)
- One ⁵⁄₁₆" × 2" × 3¼" U-bolt
- Two 2-foot lengths ¹⁄₁₆" wire rope
- Four ¹⁄₁₆" ferrule and stop sets
- Eight Simpson Strong-Tie GA2 3¼" × 1¼" gusset angles
- One 5 × 8-foot piece EPDM pond liner
- Two kitchen sprayer hose guides (Danco, model 80956)
- One small tube silicone caulk
- Two ½"-diameter × 6" copper sweat repair couplings
- One 35½" × 79" × 2"-thick sheet florist foam
- One 3 × 7-foot piece landscape fabric
- One quart natural deck sealer or wood waterproofer (optional; see step 1)

SPECIALTY TOOLS

- Copper pipe cutter
- Pneumatic staple gun
- Two 36" bar clamps
- Two 18" bar clamps

Staple with Ease

A pneumatic staple gun is ideal for this project, and it's used to drive ¾" and 1¼" staples. Alternatively, you can use a manual staple gun for all ¾" staples and substitute 1¼" #8 deck screws for the 1¼" staples.

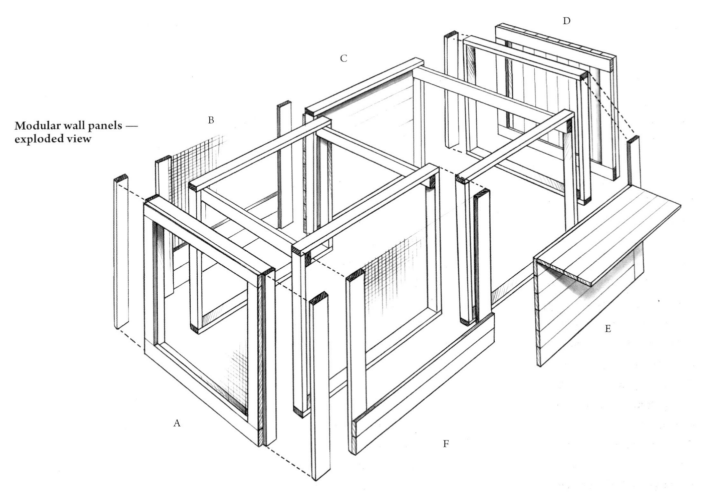

Modular wall panels — exploded view

A

B

C

D

E

F

1. CUT AND PAINT THE MAIN PARTS.

Following the modular concept, you will cut most of the coop parts to size, then paint them (as applicable) before assembling the main structure. You can use any paint or other suitable finish you like. The original house has Behr 400B-5, Grape Green, Flat Matte for its yellow paint and Behr 150B-7, Poinsettia, Flat Matte for its red paint.

Using a miter saw, square off one end of each piece of lumber, then cut it according to the lengths specified in the cutting list at right. After all of the pieces are cut, prime and paint them according to the color indicated (or substitute a color of your choice).

NOTES:

> *It will be difficult to tell the difference between the rot-resistant and common 2×4s after they are painted. Keep the two types of pieces separated during cutting and painting, and mark the bottom of each rot-resistant piece with an x after the paint is dry.*

> *Do not remove more than ½" when squaring off the cedar fencing. Sealing these pieces is optional. They will naturally patina to a grayish color if left unsealed. Alternatively, you can seal all surfaces of the cedar pieces before assembly and apply a wood sealer to the exposed faces yearly to maintain an unweathered look.*

CUTTING LIST

Pieces in the list marked with an asterisk (*) are easiest to cut if you set up a stop block that is square to your saw blade for accurate, consistent cuts (see page 84).

Materials	Cut pieces	Paint yellow	Paint red
Two 10-foot rot-resistant 2×4s	Six at 35½"*	✖	
Seven 10-foot 2×4s	Twenty-one at 35½"*	✖	
Three 8-foot 2×2s	Two at 54"		✖
	One at 35½"*	✖	
	Two at 24"		✖
	Four at 9"		✖
Three 8-foot 1×12s	Two at 80½"		✖
	Two at 36"		✖
Seven 8-foot 1×4s	Seven at 35½"*	✖	
	One at 32½"	✖	
	Four at 28½"	✖	
	Two at 17"	✖	
	Six at 12"	✖	
Three 8-foot 1×2s	Two at 81½"		✖
	Two at 35½"*		✖
	Five at 8"	✖	
Twenty-one cedar pickets	Thirty-eight at 35½"*		
	Four at 28½"		
One 4 × 8-foot sheet ½" plywood	One at 36" × 79" (apply wood sealer to all surfaces)		
	One at 14" × 34"	✖	
	One at 8" × 28"	✖	
12"-diameter concrete tube form	One at 12"-long	✖	
Hardware cloth	One at 33" × 33"		
	Two at 35" × 38"		

2. ASSEMBLE THE WALL PANEL FRAMES.

Build the basic lumber frames for the six wall panels, using four 35½" painted 2×4s for each frame. Position one rot-resistant piece at the bottom with two common pieces for the sides and one for the top. The sides butt against the inside faces of the top and bottom so all pieces are flush at the outside of the frame. Assemble each frame by drilling ⅛" pilot holes and driving two 3" deck screws through the top and bottom pieces and into the ends of the side pieces. Check all mating pieces and the assembled frames for square as you work.

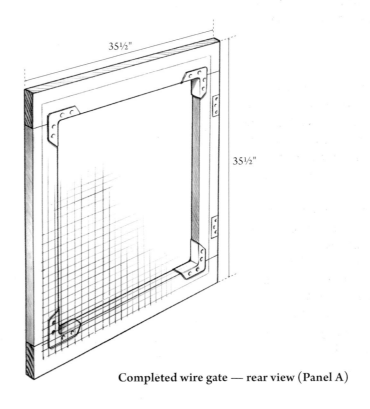

common 2×4

Assembled wall frame

rot resistant 2×4

3. BUILD THE WIRE GATE (FOR PANEL A).

Lay out a frame with two 35½" 1×4s for the top and bottom pieces and two 28½" 1×4s to fit in between for the sides. Apply glue to the ends of the sides and clamp the frame together. Reinforce the corner joints with Simpson Strong-Tie RTA12 corner angles and 32¾" #8 pan-head screws.

On one side of the frame (which will be the back side), install two 3" hinges, one 6" from the top and the other 6" from the bottom of the frame. Center a 33" × 33" piece of hardware cloth over the same side of the frame and clamp it in place. Starting at one corner and working your way around the frame, staple the hardware cloth to the frame pieces every 4" to 6", using ¾" staples.

NOTE: *You will attach this gate to wall panel A after the coop is assembled and in place.*

35½"

35½"

Completed wire gate — rear view (Panel A)

4. COMPLETE PANELS B AND F.

Lay down one of the 2×4 wall frames on the ground. Center a 35" × 38" piece of hardware cloth over the frame and clamp it to keep it taut. Starting at one corner and working your way around the frame, staple the hardware cloth to the frame every 4" to 6" with ¾" staples.

Mark each side piece of the frame 2" from the top of the frame, on the narrow edge of the side piece. Position one of the 28½" cedar pickets flat against the narrow edge of each side piece so its top end is on the 2" mark and its outside edge is flush with the outside of the frame. Fasten the picket to the frame using 1¼" staples, driving a pair of staples near the top end, middle, and bottom end of each board.

Position two 35½" cedar pickets horizontally along the bottom of the frame, below the vertical pickets, and centered over the frame. The last picket should overlap ½" onto the bottom (rot-resistant) 2×4 of the frame. Staple the bottom pickets with two 1¼" staples at each end. Repeat the same process to complete the second panel.

NOTE: *For best appearance, use some kind of guide (a board, straightedge, or chalk lines) to keep the staples aligned vertically. The stapling pattern will begin to show as the cedar weathers, so it looks best if the stapling is consistent among all of the finished panels.*

5. COMPLETE PANEL C.

Lay down one of the wall frames and mark both side pieces 2" from the top of the frame. Place one 35½" cedar picket horizontally so its long top edge is on the 2" marks. Arrange nine more pickets below the first so that all ends are flush with the outsides of the frame. Adjust the spacing between pickets, if necessary, so the bottom picket overlaps about ½" onto the bottom (rot-resistant) 2×4 of the frame. Staple the pickets to the frame with a pair of 1¼" staples at each end.

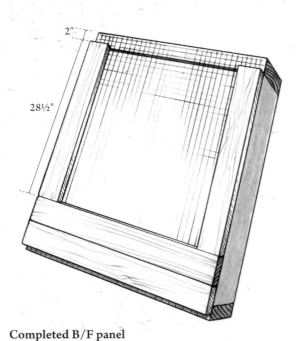

Completed B/F panel

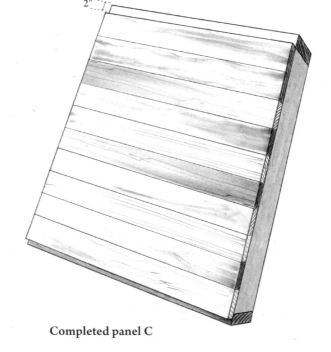

Completed panel C

6. BUILD THE WOOD GATE (FOR PANEL D).

The wood gate starts with a 1×4 frame that's identical to the wire gate frame you built in step 3. Build the frame, then install the 3" hinges 6" from the top and bottom, as with the wire gate.

Lay the frame face up (hinge side down) and arrange ten 35½" cedar pickets vertically so all are flush with the top and bottom of the frame and are spaced evenly (as needed) across the frame's width. The two outside pickets should be flush with the outsides of the frame. Staple the pickets with a pair of 1¼" staples at each end.

NOTE: *You will attach this gate to panel D after the coop is assembled and in place.*

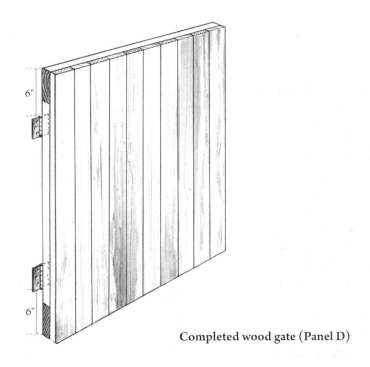

6"

6"

Completed wood gate (Panel D)

7. BUILD PANEL E.

Panel E is similar to panel C with its horizontal cedar pickets (see photo on page 135), but it features a convenient hatch door for collecting eggs. Start by laying out a wall frame and ten cedar pickets, just as you did with panel C, in step 5, but do not fasten the pickets yet. Insert two 3" hinges between the fourth and fifth pickets from the top to make sure the gap between the two pieces will provide enough space for the hinges. The first picket should not move past the 2" mark at the top of the frame. All pickets will move down. Fasten the fifth picket and the remaining pickets below with a pair of 1¼" staples at each end (the top four pickets will be used for the egg hatch door).

Clamp a 32½" 1×4 to the back side (interior side) of the fifth picket from the top of the frame assembly. Staple through the front of the picket with six 1¼" staples spaced evenly along the board's length. The 1×4 will serve as backing for mounting the hatch door hinges.

Build a frame for the hatch door with 17" 1×4s for the top and bottom pieces and 12" 1×4s fit to the sides. Join the frame pieces with 3-surface corner braces, using the included screws.

Trim ¼" from the length of the four remaining pickets (to provide a clearance gap for the hatch door). Lay the hatch door frame face up on a worktable and center one of the pickets over the frame with its bottom edge aligned with the bottom of the frame. Fasten the picket with a pair of 1¼" staples near each end

of the frame, driving the staples through the front of the picket. Install two more pickets above the first. Then, install the final picket, using two staples placed close together near each end of the door frame and add four more evenly spaced staples running horizontally along the top piece of the door frame (this picket requires secure fastening because it overhangs the door frame).

Hang the hatch door to the wall panel, using the 3" hinges. Install a U-bolt onto the top picket of the door to create a handle; secure the bolt with nuts on the front side of the picket and a uni-washer and nuts on the back side (the washer and two nuts come with the U-bolt; you provide the additional two nuts).

Add wire rope stoppers to the hatch door: Screw a #10 screw eyelet into the top left side of the hatch door frame. Screw a corresponding #10 screw eyelet into the side of the 2×4 wall frame. Loop a 1⁄16" wire rope into the wall-frame eyelet and secure it with a 1⁄16" ferrule, applying moderate pressure with wire cutters. Hold the hatch door open at a 90-degree angle. Loop the loose end of the wire rope to the door's screw eyelet and secure it with a ferrule. Repeat the process to install a stopper at the other end of the hatch door.

Install a 2" barrel-bolt latch at the top right side of the hatch door.

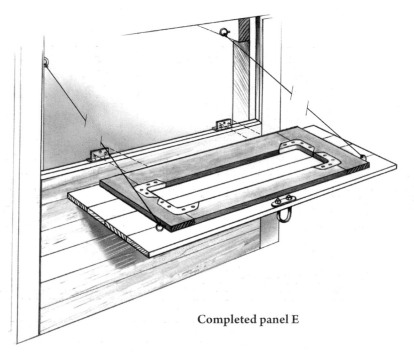

Completed panel E

8. ASSEMBLE THE WALL PANELS.

Clamp wall panels B and C together, applying the clamps to the two mating 2×4s that will be at the center of the coop structure. Make sure all edges of the mating pieces are flush and the assembly is square and level. Mark the inside face of one of the clamped 2×4s, 6" from its top and bottom ends. Drill a ⁵⁄₁₆" hole through both 2×4s at each mark, then anchor the panels together with two ⁵⁄₁₆" × 3½" hex bolts with washers and nuts. Repeat the same process to join panels E and F.

Position panel A over the end of panel B so the side of A is flush with the outside face of B. Make sure the frames are square (horizontally and vertically) and clamp them together at the corner.

Mark on the inside face of panel A's side piece, 6" from the top and bottom of the board. Secure the panels together with two Simpson GA2 gusset angles fastened with twelve 1¼" deck screws, placing the angles at the 6" marks, as shown in the *Panel corner detail*. Use the same method to attach panel A to F, panel C to D, and panel D to E.

Install three 35½" 2×4 top braces to increase the rigidity of the frame assembly, as shown in *Assembled coop structure*. Clamp each of the pieces in place, drill ⅛" diagonally opposed pilot holes through the brace and into the wall frame member, and fasten the brace with a pair of 3" deck screws at each end.

Panel corner detail

Assembled coop structure

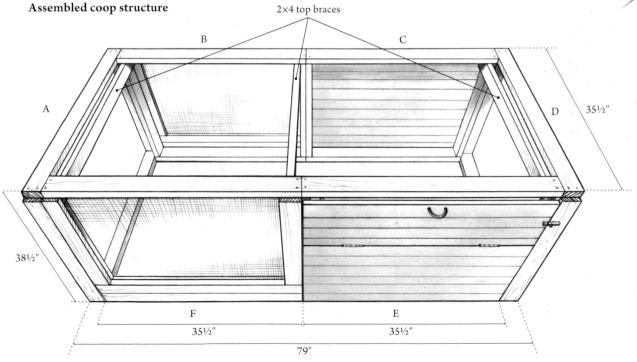

2×4 top braces

9. BUILD THE GARDEN ROOF BOX.

The roof box is a rectangular frame of 1×12s and has vertical 2×2 blocks to reinforce its corner joints and horizontal 2×2 ledges that rest on top of the coop below.

To prepare the end pieces of the box, glue and clamp a 9" 2×2 vertically to each end of a 36" 1×12 so the 2×2s are flush with the top edge and ends of the 1×12. On the same face of the 1×12, glue and clamp a 24" 2×2 horizontally so its bottom edge is 2¼" up from the bottom edge of the 1×12 and it's centered side-to-side. Make a mark 2" from each end of the two horizontal 2×2s. Drill a ⅛" pilot hole at each mark, and fasten the 2×2s with 2" deck screws driven through the 2×2s and into the 1×12. Repeat the same process to prepare the other end piece for the box. Secure the two 9" 2×2s to the 1×12 by screwing two 2" screws at 1" and 7" down on each corner piece.

Prepare the box side pieces with 54" 2×2 ledges and 80½" 1×12s. Center the 2×2s side-to-side on the 1×12s, 2¼" up from the bottom edges, and fasten them with glue and screws. Drive a screw 2" and 18" from each end of the 2×2s.

Assemble the garden box by laying the 36" × 79" sealed ½" plywood piece on top of the assembled coop structure, aligning the outside edges. Set an end and side box piece onto the plywood so the horizontal 2×2s are resting on the plywood and the 1×12s are flush at the corner. The side piece should fit over the end of the end piece. Clamp the pieces together at the vertical 2×2 block. Fit and clamp the remaining two box pieces in the same manner so that all four corners are clamped.

At both ends of each end piece, make marks at 2" and 6" down from the top and 1½" in from the corner. At the ends of the side pieces, make marks at 4" and 8" down from the top and 1½" from the corner. Drill ⅝" holes at the marked locations, making sure the holes go through straight. Secure the corner joints with ⁵⁄₁₆" × 3" galvanized carriage bolts, using a washer and nut on the inside of the box.

Drill two holes for drain spouts: Make a mark 7" down from the top and 6" in from the corner of the box frame, marking the side 1×12. Do the same thing on the other side piece directly across from the first marking. Drill ¾" holes at each marked location.

NOTE: *It's best to locate your drain spouts above panels B and F so water doesn't drip on you while you're collecting eggs on the other end (panels C and E).*

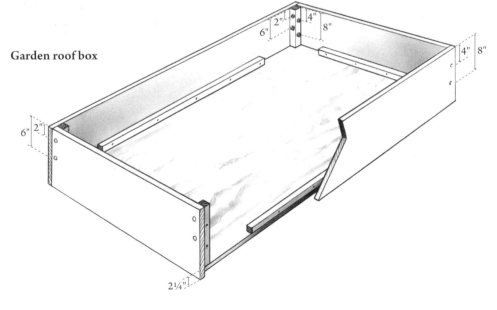

Garden roof box

10. ADD THE WATERPROOF LINER.

Center a 5 × 8-foot piece of EPDM pond liner within the garden roof box. Beginning at one corner and working your way around, secure the liner to the top edges of the 1×12 frame using ¾" staples, spacing the staples every 3" to 4". The liner should be a little loose — not taut — on the inside of the box. Make a triangle fold with the liner at the corners and secure it in place with ¾" staples. Trim off any excess liner so it's flush with the outside edge of the roof box, using a utility knife. Don't staple the pond liner above the two drain spout holes, as you will need access behind the pond liner to install the drains.

11. CREATE THE DRAIN SPOUTS.

At each of the two locations of the drain holes, mark where the pond liner meets the drain holes and cut a small X with a utility knife. Lift up the pond liner and insert the male end of a kitchen sprayer hose guide adapter through the outside of the liner, with the threaded end pointing to the inside. Add the adapter's retaining nut onto the threaded male shaft on the inside of the liner and tighten to create a watertight seal. Set the liner back into place and finish stapling it to the roof box above the two drain spout holes.

From the outside of the roof box, apply a thin bead of silicone to the interior of the hose guide adapter through the hole in the 1×12 roof box. Insert a ½" copper coupling through the hole in the box and into the adapter. Do not insert the copper farther than the interior end of the adapter.

NOTE: *It's okay if you can't find the specified model of hose guide adapter. The main criteria is that the inside dimension fits the* ½" *copper piping snugly (approximately ¹¹⁄₁₆" for the inner diameter). The length of the adapter's threaded shaft is not important (the one for this coop is about 1¾" long).*

Spray hose adapter installed on garden box interior

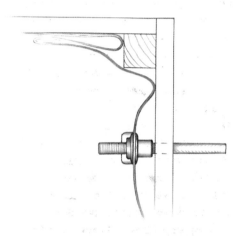

Pond liner corner fold

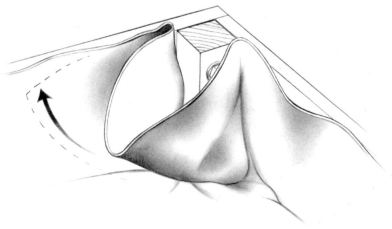

12. ADD THE ROOF TRIM.

Center a 35½" 1×2 on the top edge of one of the roof box ends so it overhangs the outside face of the 1×12 by ½". Fasten this trim piece to the 1×12 with 1¼" staples driven every 4" to 6".

Fit an 81½" piece of 1×2 trim along one of the roof box's long sides, butting its inside edge against the end of the installed end trim. Fasten the long trim piece as with the end trim. Install two more trim pieces along the remaining two roof box edges, using the same technique.

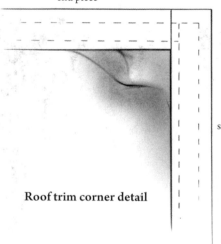

end piece

side piece

Roof trim corner detail

13. BUILD THE NESTING BOX PLATFORM.

Install the 1×4 support rails for the nesting box platform: Inside the coop, make a mark 19" from the ground onto each vertical 2×4 of wall panels C and E. Make similar marks onto the vertical 2×4s of panels B and F, but mark only on the 2×4s that meet panels C and E. Align the top edge of one 35½" 1×4 with the marks on panel C, and clamp it in place. Fasten this rail with two 1¼" staples at each end. Install another rail in the same way on frame E, then add a third rail at the marked location between frames B and F.

Assemble the nesting box and its platform: Arrange four 12" 1×4s set on edge to form a box frame, with two parallel pieces fitting over the ends of the other two pieces so the frame measures roughly 12" × 13½". Assemble the frame with glue and 1¼" staples, using two staples at each joint. Set the nesting box on top of the 14" × 34" piece of ½" plywood so the box is 2" from one end of the plywood. Carefully flip the pieces over. Drill two 1/8" pilot holes through the plywood and into the box frame below, and fasten the plywood with two 1¼" deck screws.

Position the platform and box right side up onto the 1×4 support rails attached to frames C and E, toward frame D. Drill two 1/8" pilot holes and drive 1¼" deck screws through the plywood and into the top edge of the rail on panel C. Do the same to fasten the platform to the rail on panel E. Set the 12" piece of concrete tube form into the nesting box frame to complete the nesting box.

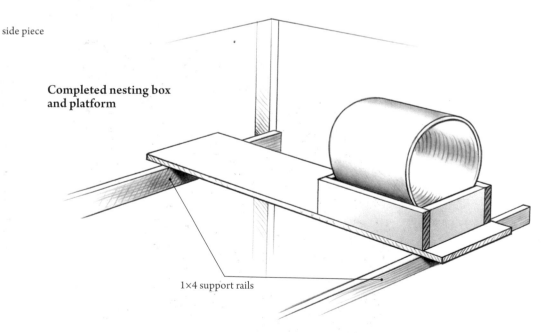

Completed nesting box and platform

1×4 support rails

14. ADD THE LADDER AND ROOSTING BAR.

To build the ladder, arrange five 8" 1×2s evenly along the length of the 8" × 28" plywood piece, centering the 1×2 rungs across the plywood's width. Glue and clamp the rungs in place, then fasten them to the plywood with one 1¼" staple at each end.

Drill two ⅛" pilot holes into the top edge of the top 1×2 rung, each 1" from an end. Install #10 screw hooks in the holes. Align the ladder with the edge of the nesting platform, and mark two corresponding locations where the eye hooks meet the platform. Drill two ⅛" pilot holes into the edge of the plywood and install two #10 screw eyelets in the holes. Once the coop is set in its final location (step 15), simply hook the ladder onto the eyelets to install it.

NOTE: *The chickens will need the ladder only until they're old enough to hop onto the nesting platform. When that time comes you can remove the ladder, if desired, to create space for an additional roosting bar.*

To install the roosting bar, position the 35½" 2×2 under the nesting platform, roughly centered on the coop's width (about 15" from panel E). Align the end of the roosting bar to the 1×4 support rail between panels B and F. Drill a pilot hole and drive a screw through the rail and into the end of the roosting bar. Drill a ⅛" pilot hole into the plywood of the nesting platform and into the roosting bar below, and fasten through the plywood with a 1¼" deck screw.

Ladder and roosting bar installed

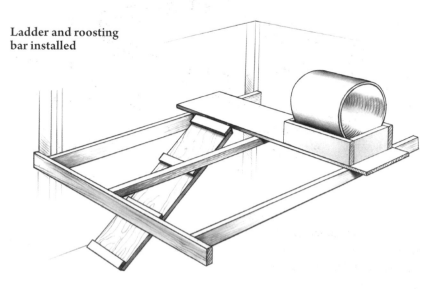

15. HANG THE DOORS.

Place the chicken coop in its final location, making sure it is on solid, level ground. Hang the wire gate onto the panel A frame and the wood gate onto panel D. Position the gates ½" below the bottom of the roof box and centered over the panel frames. Mount the hinges to the outside edges of the panel frames. Install a hook-and-eye latch at the midpoint of each gate, with the hook attached to the gate and the eye mounted to the adjacent coop panel.

16. PLANT YOUR GARDEN.

The garden roof is essentially a raised garden bed. (It just happens to be on a chicken coop!) Therefore, one of your main goals is to retain moisture while eliminating excess water. Inverted nursery trays do a great job of creating a drainage layer that lets water drain through the soil and exit through the drain spouts.

Cover the floor of the roof garden with 2"-thick sheets of florist foam (or 2"-thick plant-safe Styrofoam). Cover the foam with a piece of landscape fabric; this will help keep the soil from washing down the drain. Now you can fill your garden bed with soil. Because raised garden beds can dry out quickly, it's *very* important to use potting soil as your main soil component, and mix in about 20 percent of compost for some organic matter. Potting soil retains moisture well, and it's lightweight — you don't want to load the roof with too much weight.

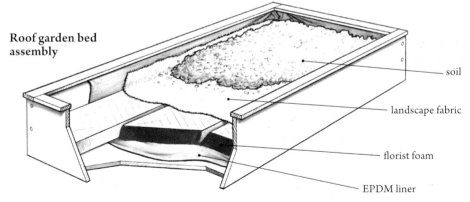

Roof garden bed assembly

soil

landscape fabric

florist foam

EPDM liner

Once you've added your plants, treat your roof garden like any other raised garden bed: Water when the soil feels dry, and mix in composted chicken manure to help your plants thrive.

PALLET COOP

Novella Carpenter is an urban farming legend. Her books *Farm City* and *The Essential Urban Farmer* tell of her experiences raising animals and vegetables on an empty lot in Oakland, California. She approached us with the need for a bigger chicken coop to house her growing flock. Her healthy and productive farm is the product of a smart, scrappy, and resourceful woman: recycled wire mesh, scavenged benches, an outdoor kitchen, and old corrugated metal roofing dot the landscape. In keeping with the values and aesthetics of the farm, our design uses surplus shipping pallets as the building blocks of the coop. Novella responded with two thumbs-up and artisanal snacks throughout the on-site build.

The coop features a cantilevered nesting box and roosting area, under which is a dry storage area. This coop is one of our largest, measuring about 100 square feet in area, so this would be a good choice if you want a lot of chickens or a coop where the chickens would not need a separate run.

We imagined the pallets as individual wall components, ready to be combined to form a small building that would look like more than just the sum of its parts. To achieve this, we spent a day disassembling the pallets and cleaning them up. We carefully removed the slats from each pallet and ran the slats through a bench-top planer to remove the top layer of paint and grit (this can also be done with a belt sander, orbital sander, or palm sander). On some of the slats we left behind a hint of the pallet's blue patina to remind the chickens of their coop's humble origins.

Whether you scavenge or buy the shipping pallets, be sure they are all the same length, width, and thickness. This will make the building process much easier. Also, you can leave the pallets intact to save yourself some time; however, since the slats we removed became the siding for other parts of the coop, you will need to calculate and purchase additional siding.

A note about the difficulty level: This coop does not require advanced building skills to construct, but because the main structure is based on nonstandard materials (namely, the pallets), there's a good chance you'll have to adapt the design shown here to accommodate your own materials and site conditions. For the same reason, we recommend that you have a working knowledge of basic frame construction, or find a handy helper who does.

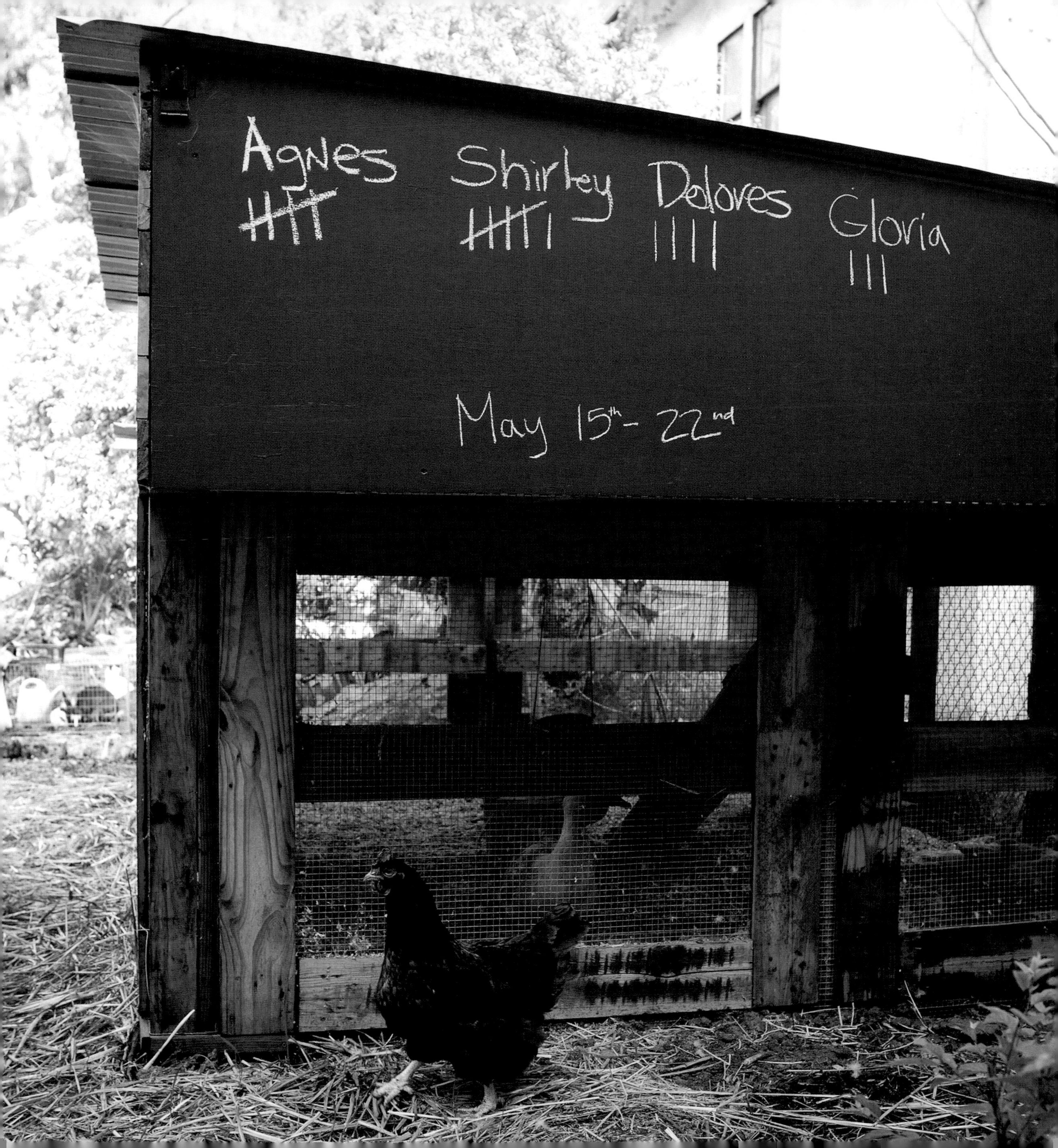
Agnes
Shirley
Delores
Gloria
May 15th - 22nd

MATERIALS

- Eight shipping pallets (40" × 48" for project as shown)
- Two 12-foot 2×4s
- Thirty 10-foot 2×4s
- Twenty 8-foot 2×4s
- Two 12-foot rot-resistant 2×4s
- Two 10-foot rot-resistant 2×4s
- Two 10-foot 2×2s
- Two 4×8-foot sheets ¾" plywood
- Five hundred 2½" deck screws
- Four hundred fifty 1¼" deck screws
- One hundred 1½" roofing screws with neoprene washers
- 50 linear feet ½" galvanized hardware cloth, 18-gauge, 48" wide

- Four ground stakes (scrap wood or rebar)
- Ten 8" × 8" × 16" (nominal dimensions) concrete blocks
- Twenty plastic zip ties
- 1×6 cedar siding (for 40 square feet of coverage) or three additional shipping pallets
- One exterior door with hinges (door shown measures 30" × 70")
- Two 48"-long piano hinges with screws
- Three 3" exterior hinges with screws
- Four exterior hasp latches with screws

- Eight 2 × 10-foot sheets corrugated roofing
- Four 8-foot pieces wiggle board (matched to roofing material)
- One gallon exterior clear finish
- One pint chalkboard paint

SPECIALTY TOOLS

- Cat's paw
- Crowbar
- Planer or orbital sander and sandpaper (80-, 120-, 150-, 220-grit)
- Mason's string

1. TAKE APART THE PALLETS.

Gather your eight pallets from the sources of your choice (see The Mighty Shipping Pallet, on page 148, for tips on finding pallets and the best kind to use for chicken coops). We prefer pallets that have a solid structure and wood slats nailed or screwed on top, as shown.

Remove the top slats on the pallets by pulling the nails, screws, or staples with a cat's paw, crowbar, and/or any other tool that gets the job done efficiently. Be careful not to damage the slats because you will use them later as siding for the coop. If the pallet frames are unstable after you remove the slats, drive nails or screws into the corners of the frames, as needed, to make them structurally stable.

Run the removed slats through a benchtop planer, or use a power sander, to gently clean up one face of each piece. If you're using a planer, start slowly and take off only ¹⁄₃₂" with each pass; taking off more can damage your tool. Run all the slats through once, then lower the planer according to the machine's directions, and run them through again. Repeat if desired; we left a bit of the grit and paint on our pieces.

NOTE: *While scavenging pallets, keep an eye out for a nice exterior door to use for your coop. We found one that measured 30" × 79" — too tall for our design but the perfect width. So we chopped it down to fit. Be creative!*

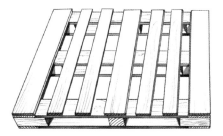

Complete pallet

Pallet with top slats removed

The Mighty Shipping Pallet

The pallet first appeared on the shipping scene in the early 1900s and was further developed around World War II. Pallets enabled loads to be moved by forklift, to fit in shipping containers and train cars, and to stack on tall racks. To this day, universal standardization of pallet dimensioning remains elusive, with dozens of sizes existing throughout the world. The most common in the United States is the grocery pallet, measuring 48" × 40". These make up 30 percent of all new wood pallets produced in the United States.

Wood pallets must be treated to avoid carrying invasive species of insects and plant diseases. This is done in two ways, identified by a stamp on the pallet. An "HT" stamp indicates that the pallet has been heat-treated to 132.8 degrees F for at least 30 minutes. "MB" implies the pallet has been fumigated with chemicals (formerly methyl bromide, which has since been phased out, though the stamp lettering remains). The latter is less desirable for building a structure where your egg-producing chickens will live, peck, and scratch.

As builders we see the pallet as a structural element, a prefab building block to transform and combine. There are approximately two billion pallets in circulation around the planet, and about two-thirds of them are used only once. They are plentiful and often inexpensive or even free. To find pallets, ask around in industrial areas of town and at hardware stores, produce markets, and factories. Keep your eyes out when you drive around; when you look for them, pallets will begin to appear everywhere. (Now that we've been building with them, Pallet Homes, Pallet Barns, Pallet Furniture, Pallet Offices, Pallet Refugee Housing, and Pallet Pavilions are on our radar.)

Even if you work in the shipping industry, you probably won't know what has been carried, stacked, spilled, and broken on your pallets. Wear gloves whenever you work with pallets. If you will be sanding or machining pallet wood, wear safety glasses and a respirator that filters dust and fumes. Further, pallets are usually constructed to be incredibly sturdy, and thus difficult to take apart. A special type of nail-pulling tool called a cat's paw, combined with a hammer, is great for digging out recessed nails. A drill may be needed if there are screws holding the pallet together. Be gentle and take care to keep the slats in good condition, as you will reuse them as siding on your Pallet Coop.

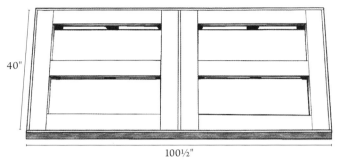

Side wall frame

40"

100½"

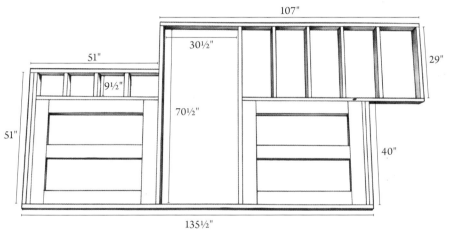

Front wall frame

107"

51"

30½"

9½"

51"

70½"

29"

40"

135½"

2. BUILD THE WALLS.

Construct the four walls using the pallets: rot-resistant 2×4s for the bottom plates, and common 2×4s for everything else. Assemble the framing with 2½" deck screws. Remember, if your pallets are sized differently from ours (40" × 48"), adjust the dimensions accordingly as you build these walls.

Build the two side walls, which are identical, using two pallets for each and framing them with lumber on the sides, top, and bottom. Also add a single vertical 2×4 between the pallets. Frame the front wall with two more pallets. Size the door opening so it is ½" wider and taller than your door (our door measures 30" × 70", so the door opening is 30½" × 70½").

Frame the rear wall, using the two remaining pallets. Note that this wall is shorter than the front wall, to create the slope of the roof. The slope should be at least 1" per linear foot. Our coop is 8 feet wide (from front to back), so our rear wall is 8" shorter than the front wall. If you live in a snowy climate, you should double the roof slope.

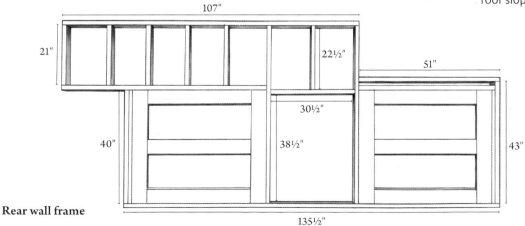

Rear wall frame

107"

21"

22½"

51"

30½"

40"

38½"

43"

135½"

Mesh for side, front, and rear walls

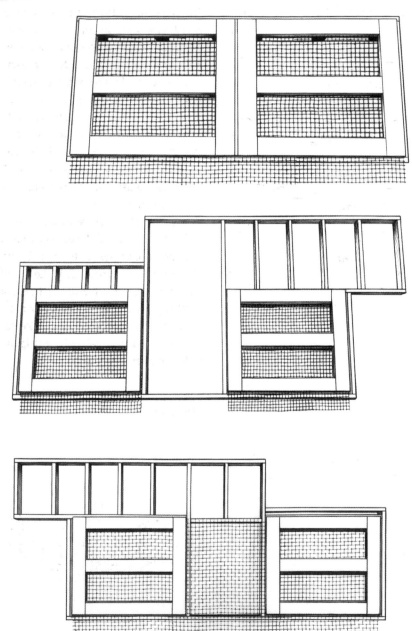

3. ADD THE HARDWARE CLOTH.

Cut pieces of hardware cloth to cover the pallet areas of each wall, leaving 8" extending below the bases of the walls; you will bury this portion into the ground to protect against burrowing predators (see Protection from Predators and Rodents, on page 16. Also see page 26 for tips on working with hardware cloth). Note that the rear wall gets a single piece of mesh across its length, while the front wall does not get mesh over the doorway. Secure the mesh using some of the planed (or sanded) pallet slats, fastening the slats to the pallet frames with at least three 1¼" screws per slat.

4. PREPARE THE SITE.

Choose an area of dry, level ground for locating your coop, and grade the soil as needed to create a flat working space. Following your wall dimensions, lay out the footprint of the assembled coop onto the site, using stakes and mason's string, then dig an 8"-deep trench along the perimeter for burying the hardware cloth (when assembled, the side walls of the coop will butt into the front and rear walls).

Arrange 10 concrete blocks so they are evenly spaced along the perimeter of the coop footprint. Excavate a small area for each block so its top face will extend a few inches above ground level. Set each block in place and check it for level. Use a long, straight 2×4 and a level to make sure all of the blocks are level with one another.

5. ASSEMBLE THE WALL STRUCTURE.

Starting with the front wall and one side wall, tip the walls up onto the concrete support blocks. If you're working alone, you can brace the walls with temporary 2×4 supports. The side wall should butt against the inside face of the front wall. Make sure the walls are evenly supported by the concrete blocks. Fasten the walls at the corner by screwing through the end of the side wall and into the front wall with 2½" screws. Raise and fasten the remaining walls in the same way.

Cut four pieces of hardware cloth, and wrap one around the bottom of each wall corner to complete the mesh barrier. Attach these mesh patches to the existing mesh on the walls, using zip ties. Cut another piece of mesh to add below the front door opening, and secure it to the front wall's bottom plate with a strip of pallet wood and 1¼" screws.

Support block layout

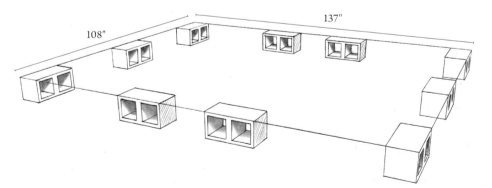

137"

108"

Securing mesh patches with zip ties

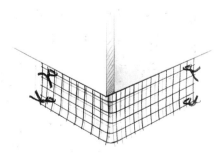

6. FRAME THE ROOF.

Cut a 2×4 to length at 100½" to extend between the front and rear walls, along the bottom of the cantilever portion of the walls; toe-screw it in place.

Cut a 2×4 rafter following the *Rafter diagram*, and test-fit it on the coop frame. At the front wall, the top edge of the rafter is flush with the top of the wall plate; at the rear wall, the rafter's top edge is ½" above the top wall plate, as shown in the *Rear-wall rafter detail*. This allows the roofing to extend over the entire rear-wall top plate. Make any necessary adjustments to the test rafter to ensure a good fit, then cut 10 more rafters to match.

Install the 11 full-size rafters with 2½" screws, toe-screwing them into the top plates and spacing them as shown in *Roof framing — top view*. Cut and install 2×4 blocking in each of the bays between the rafter pairs, following the layout in the plan. Add a short rafter that's centered over the door opening, fastening its front end to the blocking.

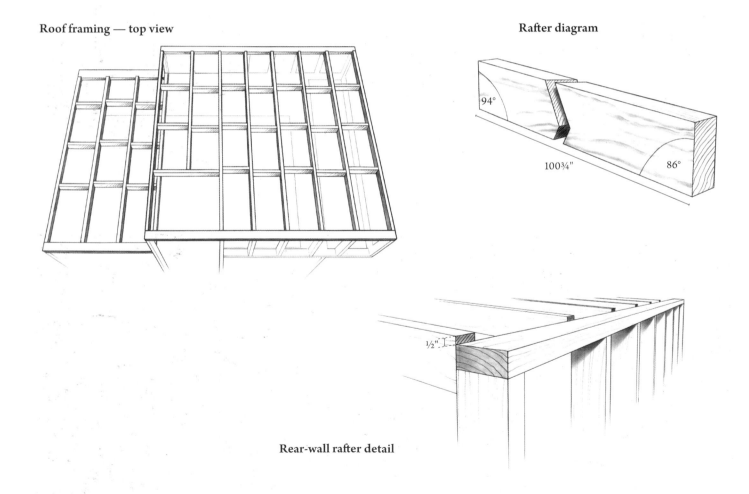

Roof framing — top view

Rafter diagram

94°

100¾"

86°

½"

Rear-wall rafter detail

7. BUILD THE NESTING BOXES.

Construct the floor for the nesting boxes, starting with a 2×2 spanning between the front and rear walls, as shown in *Nesting box floor assembly*. Cut two pieces of ¾" plywood for the floor, and notch the outside end of each piece to fit around the short studs on the front and rear walls. Install the floor pieces with 1¼" screws.

Cut a 2×4 nailer to fit between the outer rafter and the horizontal 2×4 (at the bottom of the cantilever) below, angle-cutting the top end to match the rafter's slope, as shown in *Nesting box panels*. Install the nailer by toe-screwing into the rafter and 2×4 below. Cut three plywood nesting box divider panels to fit, and

install them by screwing up through the nesting box floor and rafter (see page 17 for tips on sizing nesting boxes). Also screw the frontmost panel through the nailer. Note that the dividers have different heights because they follow the slope of the roof.

Nesting box floor assembly

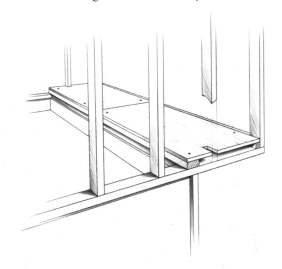

Nesting box panels

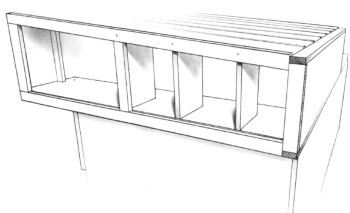

8. ADD THE ROOSTING BAR(S).

Cut a 2×2 roosting bar to fit between the front-wall stud and the frontmost nesting box panel. Install the bar with screws so it is 6" above the nesting box floor. If desired, add a diagonal 2×2 roosting bar to one of the rear corners of the coop (see page 15 for tips on sizing and installing roosting bars).

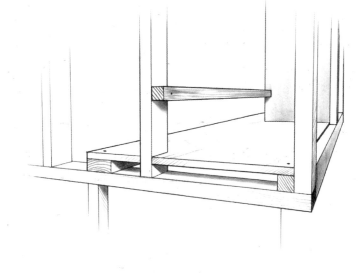

Roosting bar next to nesting boxes

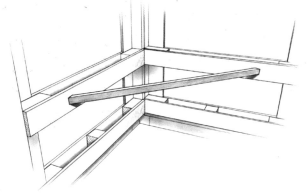

Roosting bar at coop corner

9. INSTALL THE SIDING.

Cut and install three 2×4 nailers between the upper and lower rafters at the step-down of the two roof levels, as shown in *Siding nailers*. Install siding on the upper portions of all of the walls, except for the right side wall (with the nesting boxes and roosting bar), using the planed/sanded pallet slats and 1¼" screws. Because you may run out of slats (we did), start with the side(s) of the coop that will be the most visible, then fill in any remaining areas with 1×6 decking, or create more siding boards from additional pallets.

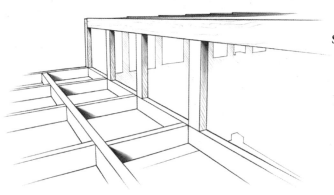

Siding nailers

Siding — front view

Siding — rear view

10. BUILD AND HANG THE DOORS.

The roosting area is covered by a hinged plywood door that's painted with chalkboard paint, providing a handy way to keep track of the hens' egg production. Both the roosting and nesting area doors are attached at the bottom with piano hinges so they swing down to open. Note how both doors are built to follow the slope of the roof.

Cut a plywood door to fit over the roosting area on the front half of the side wall. If desired, notch the upper left and right corners of the door panel to accommodate a hasp, as shown in the *Roosting door latch detail*. Install the door onto the coop framing with a piano hinge along the bottom. Paint the front face of the door with chalkboard paint, following the manufacturer's directions.

Create the door for the nesting box using leftover siding pieces. Cut the pieces to length (and width, as needed) then lay them flat with their edges together and their ends flush. Join the pieces with three cleats cut from siding scraps, screwing through the cleats and into the inside faces of the door pieces, using 1¼" screws. Position the cleats so they won't interfere with the nesting box panels or the rafter above. Install the door with a piano hinge along the bottom, and add a hasp latch.

Cut down to size your human access door, if necessary, to fit in your opening. Be sure to leave a ¼" gap on all sides. Hang the door with three 3" hinges.

NOTE: *We built our nesting box door with three siding slats to create a rectangular shape. After installing the door, we added a triangular piece of siding above the door and a straight piece below the door. The siding is flush with the outside of the door when closed, so the upper piece is perfect for mounting the catch for the hasp latch.*

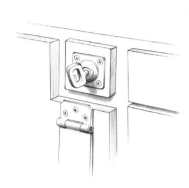

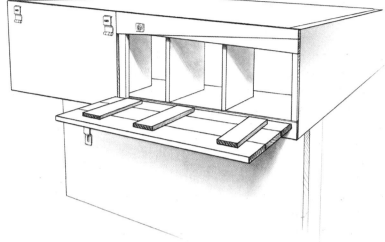

Roosting door latch detail

Roosting and nesting box doars installed

11. CREATE THE CHICKEN RAMP.

Construct the chicken ramp from leftover pieces of cedar decking or lumber. Our ramp is 84" long and has seven evenly spaced crosspieces screwed into the long support pieces. Position the completed ramp inside the coop to provide the chickens with easy access to the nesting boxes and roost, and secure its top end with screws.

12. ADD THE ROOFING

Cut and install wiggle board along the top plates of the front and rear walls; this provides support for the corrugated roofing and encloses the ends to keep out predators. Install the metal roofing, overlapping each sheet as you go, and screwing into the rafters and blocking with 1½" roofing screws (see page 26 for tips on working with corrugated roofing). Because this roof is so long and it's probably not a good idea to stand on the roof, install all of the screws for each roofing panel before adding the next, thus allowing access to the panel from the inside of the coop.

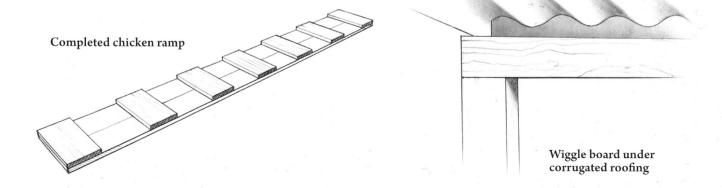

Completed chicken ramp

Wiggle board under corrugated roofing

COOPSICLE

Finally, we got to build a coop at one of our own houses. Matt moved to a new house in the middle of winter and started raising baby chicks. We wanted the coop to respond to the site, which was a steep incline. To address this, we would have to level out the ground, build with the slope, or do something entirely different. Naturally, we chose to do something entirely different.

The Coopsicle was born partly out of the opportunity to experiment on our own land. If there was ever a time to try out things that may fail, it was here, and so we decided to be ambitious. It was also born from Matt's squashed childhood dream of building a treehouse.

This time around, we decided to build up. Using a 4×4 redwood post sunk deep in concrete, we wanted the coop to rise almost magically and to use the central column for its circulation: a spiral staircase. The coop was a crash course in structural engineering, for which we sought out qualified experts, the main question being "How can we support a structure using just one central column?"

We also knew that we wanted two sides to have glass facades, not only to let in light, but also to give the coop a sense of movement. Further, the chickens would be framed in, creating a sort of lens through which to view them, and of course for them to view us. We determined the dimensions by first sourcing some windows at a local salvage yard. We found two identical windows with no interior framework and built the two side walls around them. For the floor, we made a 2×4 frame interlocked with half-lap joints and gusseted with a sheet of ½" marine plywood. The redwood post comes through the floor framing and locks into the center rafters.

The height of the floor is 48" from the ground, allowing for easy human access and cleaning (although the lower the coop is to the ground, the more stable it will be). Doors from both sides allow for reaching into the coop, and the nesting boxes are accessible from one of the doors. The spiral staircase leads to the coop through a hatch door cut into the floor.

As an ambitious experiment, it stands proud, swaying a bit in heavy winds. If your site is exposed to high winds or other extreme weather conditions, you could always support the coop with a stouter central post, or use four posts (one at each corner), or lower the coop closer to the ground.

MATERIALS

- 1 bag coarse, compactible gravel
- 10"-diameter × 48"-long cardboard concrete tube form (Sonotube or similar)
- Three 50-pound bags concrete mix
- One 12-foot rot-resistant 4×4 post (rough cut; 4" × 4" actual dimensions)
- Two 36" × 36" salvaged windows (see step 2)

- Four 10-foot 2×4s
- Twenty 8-foot 2×2s
- Two 8-foot 1×2s
- One 4 × 8-foot sheet ½" marine plywood
- 64 board feet 4/4 × 6" Tennessee red cedar siding boards (or as desired)
- 10 square feet ½" galvanized hardware cloth, 18-gauge
- One hundred 2½" deck screws
- One hundred fifty 1½" deck screws
- 1" deck screws (quantity as needed; see step 11)
- Twelve ⅜" × 4" galvanized lag bolts with washers
- Four exterior 2" × ¾" hinges with screws

- Two exterior hasp latches with screws
- One 4 × 7-foot sheet aluminum, 22-gauge
- Sixteen 1½" roofing screws with neoprene washers
- Two 1½ × 1½" hinges with screws
- One barrel-bolt latch with screws
- 6" L brackets (quantity as needed; see step 11)

SPECIALTY TOOLS

- Posthole digger or power auger
- Concrete mixing supplies (see step 1)
- Masonry trowel

Dealing with Torsion

Torsion was an issue to be reckoned with when building the Coopsicle. Torsion is twisting force. In this context, if you apply pressure to one corner of the coop, it will act as a big lever, causing the whole thing to twist. If your coop is subject to challenging site conditions (such as high winds or snow loads) or the coop stands fairly high above the ground, you might need to reinforce the post with thick metal plates (such as 3½"-wide plates bolted to two opposing faces of the post, along the first 2 feet or so from the ground) or use a 6×6 post, which would require altering the floor framing to fit the larger post.

1. PREPARE THE SITE AND SET THE POST.

Dig a 12"-diameter × 42"-deep hole, making sure the sides of the hole are relatively plumb as you work. Add a 3" to 4" layer of gravel to the hole and tamp it thoroughly with a scrap 2×4 or the 4×4 post that will support the coop. Cut a 12"-diameter concrete tube form to length so it will extend from the gravel base in the hole up to about 2" above ground level. Set the form in the hole, and backfill around it with soil, tamping the soil and checking the tube for plumb as you go.

Stand a 12-foot rot-resistant 4×4 post in the hole, position it perfectly plumb in both directions, and brace it securely with 1×2 cross-bracing. Note that the post we used is a rough-cut timber, so its actual dimensions are 4" × 4" (not 3½" × 3½" like standard 4×4s). Leave the post long for now; you will cut it to length later.

Mix concrete in a wheelbarrow or tub (or a power mixer, if you have one), and fill the form to the top, overfilling it slightly above the edge of the form (see page 36

for tips on working with concrete). Use a masonry trowel or similar tool to shape the top of the concrete into a dome to shed water away from the post, then let the concrete cure as directed by the manufacturer.

NOTE: *Call before you dig. Make sure there are no underground utility lines in or near the coop site before you break ground.*

2. BUILD THE FLOOR FRAME.

As mentioned earlier, the coop house is designed around the windows that enclose the ends. Ideally, the windows will have a simple outer sash frame with no interior wood framing (properly called muntins or muntin bars). They should also be identical in size, if possible. Our windows measure 36" × 36". If yours are a different size, you'll have to modify the coop dimensions to fit the windows. Also, keep in mind that old windows may have lead paint on them. Take care when scraping

off old paint or painting over it, and check local codes for instructions on safely handling and dealing with lead paint. If you're in a cold climate, you might want to look for double-paned windows for better insulation.

To frame the coop floor, cut six pieces of 2×4 to length at 39"; these will become

the *short* joists. Cut four 2×4 pieces to length at 60"; these are for the *long* joists. Notch the joists as shown in the *Short joist* and *Long joist diagrams*; see pages 32 and 34 for discussions on notching lumber and making half-lap joints. Construct the frame as shown, using two 2½" screws driven from the top into each joint.

Assembled floor frame

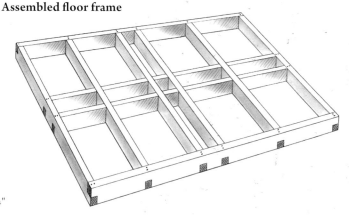

Short joist diagram

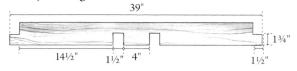

39"

14½"   1½"   4"   1½"   1¾"

Long joist diagram

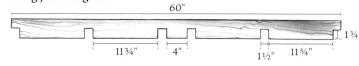

60"

11¾"   4"   1½"   11¾"   1¾"

3. ADD THE FLOOR DECK.

Cut the plywood deck to size at 39" × 60". Mark a 4⅛" × 4⅛" square hole in the center of the deck; this will accept the coop post and should be centered precisely over the square cavity in the center of the floor frame. Drill a starter hole inside the marked square, then complete the cutout with a jigsaw. Fasten the floor deck to the frame with 2½" screws.

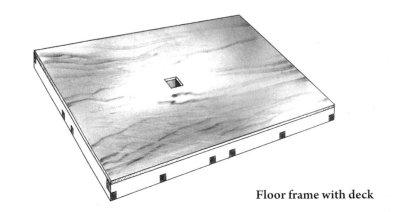

Floor frame with deck

4. INSTALL THE FLOOR.

Cut two 2×4s to length at 36". Mark the 4×4 post to indicate the bottom of the floor frame at the desired height for the coop. Keep in mind that the lower the coop sits, the more stable it will be. Center each of the 2×4s on opposite faces of the posts, with their top edges on the marks. Make sure they are level and fasten them to the post with several 2½" screws. These are temporary supports that you will remove later.

Mark all four faces of the post at 39½" from the tops of the 2×4 supports, and cut off the post with a handsaw. For those of you making modifications to the original design, the top of the post should be flush

with or just above top edges of the roof trusses' bottom chords (cross supports); see step 7.

With help from a friend or two, lift the floor frame up and onto the post until it rests on the 2×4 supports. Drill pilot holes, and anchor the floor frame to the post with ⅜" × 4" lag bolts with washers, driving three bolts into each face of the post.

5. BUILD THE WALL FRAMES.

Cut 2×2 lumber to size to create the first of the two wall frames, using a miter saw and following the dimensions and angle cuts shown in the *Wall frame diagram*. Assemble the frame with 2½" screws. Construct the second wall to match the first.

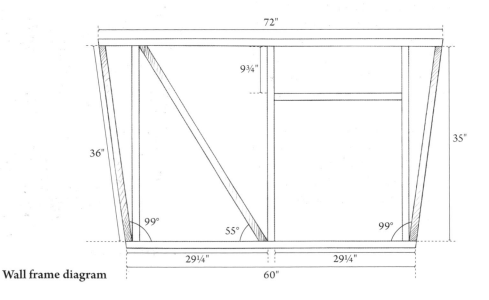

72"

9¾"

35"

36"

99°

55°

99°

29¼"

29¼"

60"

Wall frame diagram

6. BUILD THE ROOF TRUSSES.

Cut 2×2 lumber for the four roof trusses, following the *Roof truss diagram*; you can build a jig for the trusses, too, if you like. Assemble the trusses with 2½" screws.

Roof truss diagram

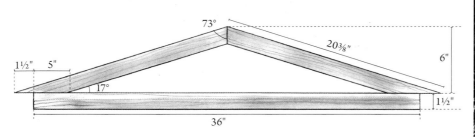

7. INSTALL THE WALL AND ROOF FRAMES.

Install two of the trusses to either side of the coop post so they are centered on the post and their bottom chords are flush with (or just below) the top end of the post. Fasten the trusses with 2½" screws.

Set the walls in place, slipping their top plates under the square recesses of the roof trusses. You may need to adjust the height of the roof trusses to get a tight fit. Note that the walls are oriented so that the openings without cross bracing are at opposing corners of the coop. Also screw the walls to the coop floor.

Install the final two roof trusses at the ends of the window walls. Install the windows by screwing through their sash frames and into the wall frames. The wall assembly should be quite stable at this point.

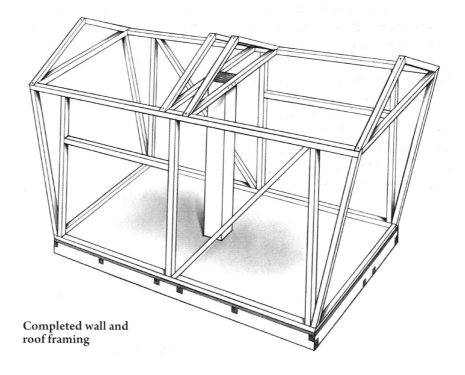

Completed wall and roof framing

8. CONSTRUCT THE NESTING BOXES.

Create the nesting boxes with plywood, following the *Nesting boxes diagram* or your own design. We built our box unit with dado joints and screws for added strength. Install the boxes onto the vertical 2×2s of one of the wall frames so that the floor of the nesting box is several inches below the horizontal crosspiece of the wall frame; this space allows you to retrieve eggs from the doorway. Note that the nesting boxes on our coop are above the hatch door in the floor, which will be at the top of the spiral staircase (see step 11).

9. INSTALL THE MESH AND SIDING.

Enclose the triangular spaces (above the windows) of the two outer roof trusses with hardware cloth, securing it to the insides of the trusses with 1×2s and 1½" screws (see page 16).

Construct two simple door frames using 2×2s and screws, sizing the frames ½" narrower and shorter than the door openings in the walls.

Install the cedar siding horizontally along the long side walls, using 1¼" screws. For the best appearance, use one piece of siding to span the length of the side, including the door; just make sure to cut

the siding flush with the door frame and to allow for a ¼" gap all around the door. If desired, you can temporarily screw the door frames to the walls to facilitate the siding installation, then hang the doors afterward.

Install siding on the ends of the coop, creating a mitered frame around the windows, if desired.

Hang each of the doors with three 2" × ¾" hinges, then add a swivel lock or other closure to each door.

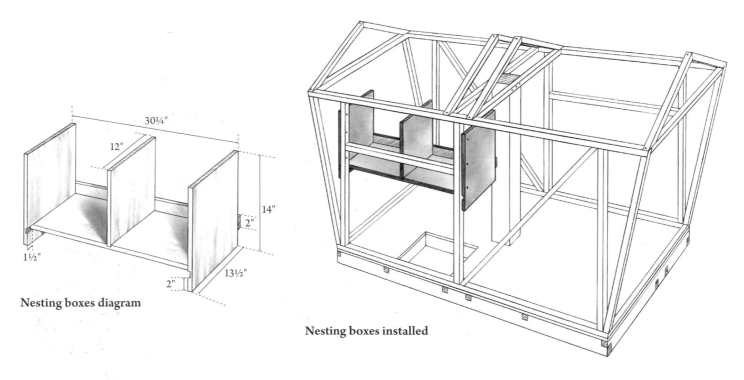

Nesting boxes diagram

Nesting boxes installed

10. ADD THE ROOF.

Using shears or an angle grinder with a cutoff disc, cut a 22-gauge sheet of aluminum to 4 × 7 feet (some sheet metal retailers will cut the material for you). Draw a line down the center of the sheet, parallel to its long edges. Mark and cut tapered "dog ears" by drawing an angled line from the middle of the sheet on each end to a point 6" from the corner, as shown.

Bend the roof along the centerline, using a homemade brake (see page 37). Test the fit on the roof trusses. The metal probably will have a little spring to it, so bend it a bit more than desired and let it spring back to its final position. Fasten the roof to the trusses with 1½" roofing screws.

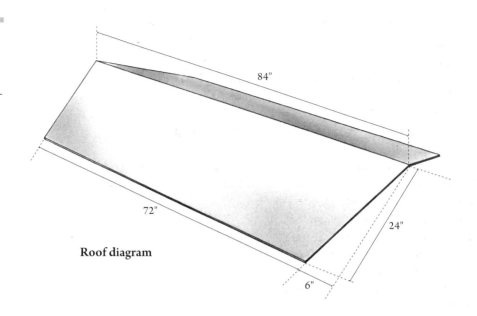

Roof diagram

11. CREATE THE HATCH DOOR AND STAIRCASE.

Remove the temporary 2×4 supports for the floor frame. Mark a square hole on the floor deck for the hatch door, staying between the floor framing members (see *Nesting boxes installed*, on facing page). Cut out the opening with a jigsaw. It's difficult to save the cutout piece for the hatch door, so cut a new piece of plywood for the door, making it a bit smaller than the opening. Install the door with two 1½" × ¾" hinges on the top side, so the door swings into the coop. Add a barrel-bolt latch opposite the hinges, also on the top side of the floor.

Cut 12"-long step treads from leftover cedar siding. You'll need one tread for every 4" of elevation from the ground to 8" below the coop's floor deck. Starting 8" below the floor at the hatch door opening, fasten 6" L brackets upside down to the post, using 2½" screws. Space the brackets 4" apart vertically, spiraling your way around and down the post. Center the brackets on the flats of the post, and chisel flat areas for the brackets at the post corners. Fasten the stair treads to the brackets with 1" deck screws driven from underneath.

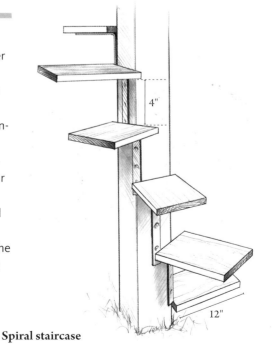

Spiral staircase

12. COMPLETE THE FINAL DETAILS.

Cut a 2×2 roosting bar to length at 48".
Position the bar 10" above the floor deck
and fasten it to the 4×4 post with at least
three 2½" screws.

Sand and finish the coop as desired.

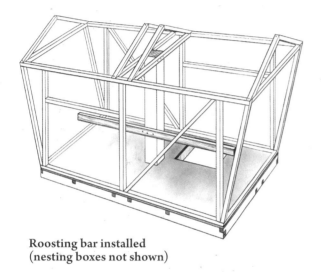

**Roosting bar installed
(nesting boxes not shown)**

The Treehouse

To imagine a treehouse is to conjure up images from child-hood, and from novels, movies, and fairy tales. The Swiss Family Robinson, the Ewoks from the Star Wars trilogy, the Hobbits in J. R. R. Tolkien's Lord of the Rings stories, and Tarzan, to name a few, all frequented these magical treetop dwellings.

More practically, treehouses have been around since ancient times as comfortable living quarters and food storage sus-pended above scavenging animals and enemies. They have been used by people in the South Pacific and Southeast Asia. In the 1700s, the English explorer Captain Cook noted an encounter with treetop dwellers in Tasmania. Today tree-houses are considered easy on the environment because no ground clearing is required to lay a foundation.

Treehouses became a political statement in northern California in the late 1990s. Protesters built and inhabited simple dwellings hundreds of feet above ground to protest the clear-cutting of 1,000-year-old redwood forests. People like Julia Butterfly Hill and Nate Madsen lived in redwoods for months and even years, dependent on supporters to deliver food but also utilizing solar panels and water collec-tion tanks to survive.

Treehouses are typically built in three ways: suspended from the upper branches, fastened directly to the tree, or supported on stilts anchored to the ground. There are at least 30 com-panies in the United States and Europe dedicated to build-ing treehouses. Special hardware, specifically a unique bolt called the Garnier Limb, has been developed to better attach treehouses directly to trees and to allow the tree to grow over time. Today a single Garnier Limb is capable of supporting 9,000 pounds.

When I was young, my brothers and I built a treehouse (with a little help from our dad). More than 20 years later it still stands, awaiting the next generation, though a few boards could use replacing. Matt, unfortunately, never realized his dream of a treehouse in urban Washington, D.C., and Denver, Colorado. He would have to wait until adulthood to build his treehouse, and unfortunately his treehouse is too small to climb into. It's for the birds, you see.

Lloyd Kahn is a northern California icon. Since the 1970s, he has been publishing resource books for natural builders and architects. On page 228 of his 1973 book *Homework*, Mr. Kahn shows a Swedish hunting cabin, with the cap-tion, "An unusual hunting lodge (shelter) from northern Sweden — Lapland. This house was used by hunters who needed a secure place to sleep away from the wolves."

When Matt and I saw this, it seemed like a great idea for a chicken coop; it is extremely predator-proof (provided you remember to close the hatch at night to keep the raccoons out) and great for a severely sloped yard where leveling is not an option. Most of the construction can happen in the shop, then the coop can be installed in place right on the post. Modern chickens are happy roosting on a piece of 2×2 screwed into their lumber and plywood home, but their ancestors roosted on actual tree branches. Building this coop was a way to span generations and give the chickens the sen-sation of living in trees again.

CUPE

The Cupe was designed as a modern composition of simple forms. Our clients live in a beautiful Midcentury Modern house in the hills of northern California and wanted a chicken coop with similar clean lines and bright colors contrasted with natural wood grain. We noted the angles of their house's low-slung roofline (pitched at 7 and 11 degrees) and inverted them for the coop's butterfly roof. The design features two cubes stacked vertically, offset in two directions, and mounted on four redwood posts.

We built half of the eight walls with MDO plywood, which has an exceptionally smooth, flat surface for painting. We then played with primary colors, mixing bright yellow, white, and red for the walls, with a bright blue roof. For the other four walls, we varnished marine plywood to accentuate the grain and provide more contrast.

The nesting boxes inside the lower level also function as stairs up to the second level, where the chickens roost. Big doors provide access to both the upper and lower levels for easy egg collection and cleaning. This coop is heavy and awkward to move, so we suggest building it as close as possible to its final location.

We had fun with the roof. It's made with a single sheet of steel, which you can bend with a homemade brake (see page 37) or have it cut and shaped at a local metal fabricator. Steel will rust over time if left untreated, so we brought our roof to a local powder-coating shop for a durable bright blue finish.

Midcentury Modernism

Modernist architecture as a cohesive movement started in Europe as a break from traditional ornamentation and is characterized by a refinement of building methods, homes with open floor plans, and an expression of technical innovation sparked by rapid industrialization. Galvanizing the movement was an ethical ideal, personified by industrial designers Charles and Ray Eames, who claimed industrialization could "get the most of the best to the greatest number of people for the least."

Architects and developers in California took such ideas and infused them with a local essence: a greater relationship between indoor and outdoor space, the hope of making design affordable to the emerging middle class, and an unpretentious simplicity of form. During the postwar years, California grew rapidly, with two out of every three residents born outside of the state, mostly settling around San Francisco and Los Angeles. Innovations from weapon and defense fabrication were assimilated into the building and construction industry, resulting in radical new design possibilities. Such growth, combined with a booming high-tech economy and a break from establishment values found on the East Coast, created fertile ground for Modernist experimentation. These conditions struck a chord with artists and designers eager to realize visions that could not be imagined anyplace else.

There was perhaps no greater advocate for applied Modernist architecture than Joseph Eichler. A developer responsible for roughly 11,000 Modernist homes, mostly in the San Francisco Bay Area, Eichler embraced the American suburban ideal with hopes that high-minded design would find its way to the "ordinary" family. He welcomed the features that illustrated the values of suburban middle-class life: the placing of the family car as a status symbol in the front of the house and "back to front" planning that situated the living and dining rooms at the back of the house, a testament to the greater value placed on privacy. Using post-and-beam or steel frame construction allowed for large openings in a house's facade for floor-to-ceiling windows, integrating garden space into daily rituals. The lack of snowfall on the West Coast also allowed for flat or very gradual rooflines that became iconic symbols of the style.

As with all movements, Modernism was not without its critics. In the late 1960s, opinion was beginning to shift, noting the coldness of pure formal expression and suspicion of any design intent that declared itself a remedy for social ills. In northern California, however, its legacy cannot be ignored. Once-shocking architectural demonstrations have become period antiques, evoking a sentimentality for times of such earnest design idealism.

MATERIALS

- Two 4 × 8-foot sheets ½" marine plywood
- Two 4 × 8-foot sheets ½" MDO (medium density overlay) plywood
- Four 8-foot 2×4s
- Four 8-foot rot-resistant 4×4s
- Fifteen 8-foot 2×2s
- Six 8-foot 1×2s
- One 5 × 5-foot sheet ¾" Baltic birch plywood (or one 4 × 8-foot sheet ¾" Euro-ply)
- Six ¼" × 3" lag screws
- Two 2" × ¾" exterior hinges with screws
- One hundred 1¼" wood screws
- One hundred 2½" wood screws
- Four 1½" roofing screws with neoprene gaskets
- 4 linear feet ½" galvanized hardware cloth, 18-gauge, 36" wide
- Exterior wood glue
- Fifty 1" brads (optional; see step 11)
- One pint primary red exterior paint
- One pint primary yellow exterior paint
- One pint primary white exterior paint
- One half-gallon exterior clear finish natural deck sealant
- Two piano hinges with screws (each approx. 30" long)
- Three 2" exterior spring-loaded hasp latches with screws
- One 4 × 8-foot sheet 18-gauge mild steel

SPECIALTY TOOLS

- Compound miter saw
- Protractor
- 48" bar clamps (at least four)

OPTIONAL SPECIALTY TOOLS

- Drill press
- ⅜" countersink drill bit
- ⅜" plug-cutting drill bit
- Nail gun
- Table saw with dado blade or router with straight bit
- Circular saw and metal-cutting blade
- Flat metal file

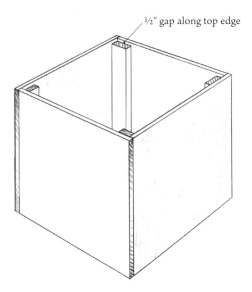

½" gap along top edge

Assembled cube — top view

1. BUILD THE CUBES.

Cut eight cube panels to size at 35½" × 35½", using ½" marine plywood for four of the pieces and ½" MDO for the other four pieces. Drill three ⅛" pilot holes along the two side edges of each panel, 1½" in from the side edges and 2" from the top and bottom edges. The face grain of the marine plywood should be vertical. If desired, use a drill press and a ⅜" countersink bit to drill a ¼"-deep counterbore for each pilot hole; you can fill these later with plugs to hide the screw heads and create a smooth, continuous surface for painting.

Cut eight pieces of 2×4 to length at 35"; these are the corner braces for assembling the cubes. Position one corner brace along the left side edge of each cube panel so it is flush with the side and bottom edges of the panel, leaving a ½" gap at the top edge. Fasten the panel to the brace with three 1¼" screws, driven through the pilot holes on the outside of the panel.

Assemble each cube by joining one panel to the next, alternating plywood and MDO, and screwing through the panels into the 2×4 braces.

2. CUT THE POSTS.

To create the 5-degree roof slope, all of the four posts are cut from rot-resistant 4×4s with a 5-degree bevel. Additionally, one plane of the roof is pitched at 7 degrees, while the other plane is pitched at 11 degrees. This means that all four posts are cut at different lengths. Make the compound cuts first, using a compound miter saw, then measure from the *longest* point of each post and mark it to length for the square bottom cut, as follows:

- **Post 1:** Angle top at 5 and 7 degrees; cut to length at 89".
- **Post 2:** Angle top at 5 and 7 degrees; cut to length at 85¾".
- **Post 3:** Angle top at 5 and 11 degrees; cut to length at 90¼".
- **Post 4:** Angle top at 5 and 11 degrees; cut to length at 87".

3. ASSEMBLE THE CUBES AND POSTS.

Remove the 2×4 brace joining two adjacent panels in one of the cubes; these will be panels A and B. Use clamps to assemble the posts and cube as shown in *Lower cube layout*. Post 1 fits inside the corner of panels A and B (where the brace has been removed). The other three posts are outside of the cube. Post 4 is not fastened to the lower cube; it will fasten to the upper cube. The bottom edges of the cube panels should be 12" above the bottom ends of the posts. Remember that the ½" gaps above the 2×4 braces are at the top of the cubes.

Make sure the assembly stands level, then fasten panels A and B to post 1 using 3" screws. Drill pilot holes on the insides of panel D and fasten through the inside of the panel and into post 2 with 3" screws. For panel C, drill three pilot holes through the corner brace and panel and into post 3, and fasten through the brace and into the post with three ¼" × 3" lag screws.

Install the upper cube in the same manner but with the cube-post orientation reversed, following the upper cube layout. Post 4 sits inside the corner of panels C and D, while post 1 is not fastened at the corner of panels A and B. Note that the materials alternate between the upper and lower cubes: on the lower cube, panel A is plywood; on the upper cube, panel A is MDO.

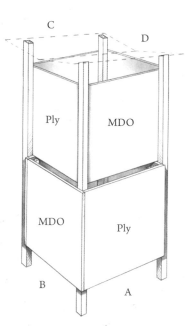

Lower cube layout (roof shown for clarity)

Upper cube layout (roof shown for clarity)

4. CUT THE ACCESS DOORS.

We chose to locate our access doors on side C of the coop. You may prefer to use a different side, depending on your site and coop orientation. However, we recommend *not* using side D for the doors because this is on the downhill side of the coop, where the rainwater will drain.

Remove the side C panel on the upper cube, and add a temporary shim to the side D panel on the lower cube (because it's not screwed to the post) to keep the lower cube intact. Cut the door from the C panel, using a table saw or circular saw, as shown in *Upper cube door cuts*. Note that the cut on the left side of the panel is straight (90 degrees), while the cut on the right side is beveled at 45 degrees. Reinstall the left and right edge pieces onto the coop, leaving off the cutout door section. Repeat the same procedure for the side C panel on the lower cube, following the dimensions shown in *Lower cube door cuts*.

Upper cube door cuts

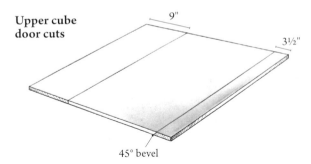

9"

3½"

45° bevel

Lower cube door cuts

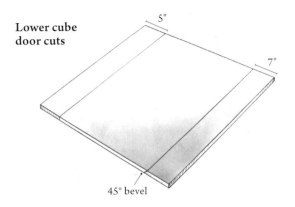

5"

7"

45° bevel

5. CUT AND INSTALL THE CHICKEN DOOR.

We chose side A — in the lower cube only — for adding a little door for the chickens to hop in and out. Again, you may want to use a different side.

Remove the side A panel on the lower cube. Mark the door cutout as shown in *Chicken door cutout*. Plunge cut with a circular saw to cut the door and finish with a handsaw; save the cutout to use as a door, if desired. Alternately, drill a starter hole inside the marked lines, and complete the cutout with a jigsaw. If desired, you can round the corners for a nice touch. Instead of doing the cut freehand, it's a good idea to clamp a straightedge or piece of lumber onto the plywood to guide the jigsaw for a clean, straight cut.

Reinstall the panel onto the coop. Use the cut piece of plywood or cut a door to size at 16" × 16" or larger from leftover ½" plywood, and hang the door with two 2" hinges on the outside of the coop, so the door swings out.

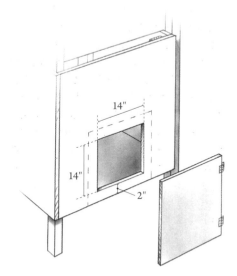

Chicken door cutout

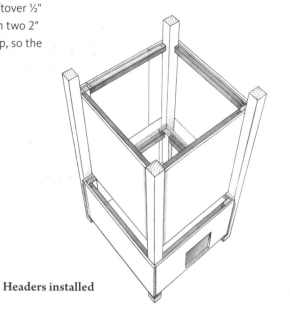

Headers installed

6. ADD THE HEADERS.

For the top of each cube, cut four 2×2 headers to fit between the corner braces and post, as shown. Position each header so it is flush with the tops of the corner braces, ½" below the top edge of the cube panel, and fasten the header into the brace or post at each end with 2½" screws toe-screwed into the lumber. Drive the screws from above or below, whichever is more convenient.

7. PREPARE THE FOOTERS, FLOOR JOISTS, AND ROOSTING BARS.

Cut six 2×2s to length at 30" and cut two 2×2s at 28". These are the footers that will be installed along the perimeter at the bottom of each cube, as shown in *Lower cube floor supports*. For each cube, notch a pair of the 30" 2×2s as shown in *Footer notching diagram* (notch four pieces total); these notches are the lower parts of a half-lap joint for installing the floor joist in each cube. Note that each cube gets two 30" notched 2×2s, one notched 12" from the end and the other at 14".

Cut four 2×2s to length at 35". Two of these will be used as roosting bars, and two will become the floor joists. Notch both ends of each of these four pieces as shown in *Floor joist notching detail*.

Install the footers on the lower cube as shown in the diagram below, toe-screwing into the post and 2×4 corner braces with 2½" screws. Set one of the floor joists into the half-lap notches, drill pilot holes, and fasten the joist to the footers with one 1¼" wood screw at each end. Repeat the same process to add the footers and joist to the upper cube.

Lower cube floor supports

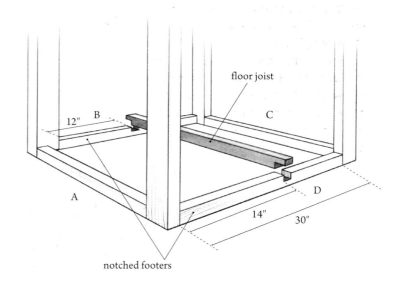

Footer notching diagram

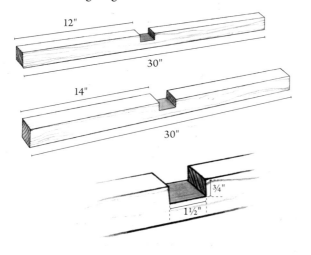

Floor joist notching detail

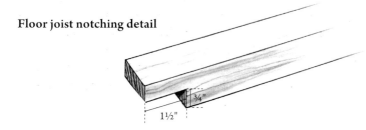

8. INSTALL THE UPPER CUBE BLOCKING AND ROOSTING BARS.

On sides C and D, where the upper cube overhangs the lower cube, cut and install additional 2×2 blocking to keep predators from entering. Toe-screw the pieces so they sit on top of the lower cube's plywood, flush with the outside edge, as shown in *Upper cube blocking installed*. This area will be covered by the upper cube's flooring.

Cut two 2×2s to length at 30" for the roosting bar supports. Install the supports on sides B and D, 6" above the footers, as shown in *Roosting bars installed*, toe-screwing into the posts and 2×4 corner braces. Position the roosting bars (the remaining 35" notched 2×2s) on top of the supports so they are 10" apart. Drill pilot holes and fasten the bars to the supports with 1¼" screws.

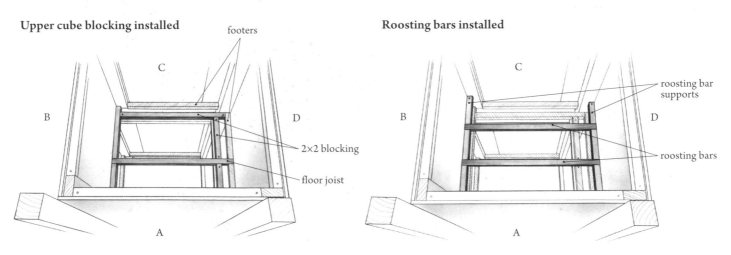

Upper cube blocking installed

footers

C

B

D

2×2 blocking

floor joist

A

Roosting bars installed

C

B

D

roosting bar supports

roosting bars

A

9. ADD THE FLOORING.

Both the upper and lower cubes get floors pieced together from leftover plywood and MDO. We used marine plywood for the lower floor and MDO for the upper floor.

For each floor, cut two outside pieces to 35" in length, using whatever width you have available. Notch the pieces to fit around the coop posts and 2×4 corner braces, as needed. Fit the outer pieces in place on top of the 2×2 footers, then fill in between with pieces cut to fit, as shown in *Lower floor — bottom view* (this floor is made up of four pieces). Fasten the floor pieces to the footers with 1¼" screws. Don't fasten the upper floor pieces yet; you will remove them to create an access hole for the chickens.

10. COVER THE MESHED OPENINGS.

Cut a piece of hardware cloth to size at 35" × 35" (see page 26). Notch one corner to fit around the internal post of the upper cube. Place the mesh on top of the 2×2 headers of the upper cube, then cut 1×2 strips to fit the inside of the cube, directly above the headers. Secure the 1×2s and mesh to the headers with 1¼" screws, as shown in *Upper cube mesh installed*.

Cut two strips of mesh to size at 5½" × 31½"; these are for enclosing the open areas between the upper and lower cubes on sides A and B, as shown in *Lower cube mesh installed*. Cut two 1×2s for each screen piece — one to go on top of the header of the lower cube and one to go on the bottom of the footer of the upper cube. Position the mesh and secure it to the headers and footers with the 1×2s and 1¼" screws.

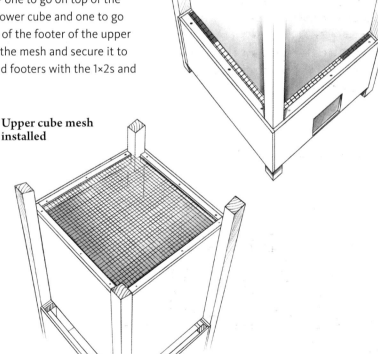

Lower cube mesh installed

Lower floor — bottom view

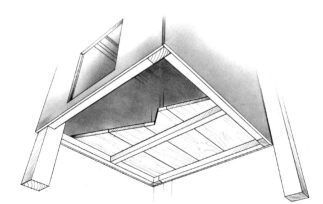

Upper cube mesh installed

11. BUILD THE NESTING BOXES.

The nesting boxes are three plywood cubes that join together in stair-step fashion to provide the chickens access to the second level. The boxes fit snugly in the coop from side to side.

NOTE: *We originally built all three cubes at the same size, as shown in the project photos, but later decided to make the middle box larger (15") than the other two (10") , as described here and as shown in* Nesting box diagram. *Also, it's better to make the combined width of the boxes a little small (by around ¹⁄₁₆") than too big, to be sure they will fit in the coop.*

Cut the following pieces from ¾" plywood:

- Two at 15" × 15"
- Two at 14¼" × 15"
- Four at 10" × 10"
- Four at 9¼" × 10"

Using a table saw and dado blade (page 32) or a router and straight bit, mill ¾"-wide × ⅜"-deep rabbets into side edges of the 15" × 15" pieces. Mill the same size of rabbet into the side edges of the four 10" × 10" pieces.

Assemble each box with wood glue and clamp it securely, checking for square. If desired, nail the rabbet joints with 1" brads (a nail gun is ideal for this) or add screws. Let the glue dry as directed by the manufacturer.

Screw the boxes to each other with 1¼" screws, following the *Nesting box diagram.* Fit the steps into the lower cube to test the fit. The top of the second box should sit about 20" above the floor of the lower cube; this is a good height for the chickens to use it as a step to hop upstairs.

Mark the general location for the access hole in the upper cube floor, above the second and third boxes. Remove the upper cube's floor pieces and mark a 10" × 12" cut at the marked location. Make the cuts with a jigsaw, then reposition the floor pieces and fasten them to the upper cube's footers with 1¼" screws.

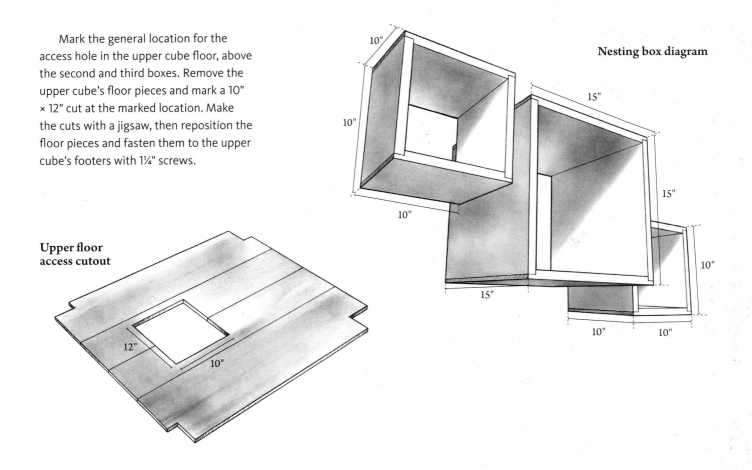

Nesting box diagram

Upper floor access cutout

12. FINISH THE COOP AND HANG THE DOORS.

Paint the MDO cube panels with white, yellow, and red paint, as we did, or choose any colors you like. For some extra touches, we painted the edges of the nesting boxes (red) and the chicken door opening (yellow). Finish the plywood panels with a natural deck sealant or other exterior clearcoat of your choice.

Install the nesting boxes, screwing from outside the coop with 1¼" screws.

Paint the large access doors (on side C), then install them with piano hinges on the left sides of the doors. Add hasp latches to each door for security.

NOTE: *If you drilled counterbores for the panel screws, cut some ⅜" wood plugs, using a plug-cutting bit. Glue the plugs over all exposed screws on the panels, pare them down with a sharp chisel, then sand the plugs flush before painting the panels.*

13. CUT AND BEND THE ROOF.

As mentioned earlier, you can cut and bend the roof yourself or have the work done by a local metal fabricator. Cut the roof to size at 48" × 60" from a single 4 × 8-foot sheet of 18-gauge mild steel, using a circular saw and metal-cutting blade. If the edges are sharp, use a flat metal file to remove any burrs.

Create a guide for bending the roof out of a scrap of plywood. Using a protractor, mark a 162-degree angle on one edge of the piece, as shown, and carefully cut it with a bandsaw or jigsaw. (The roof pitches are 7 degrees and 11 degrees, for a total of 18 degrees; 180 degrees minus 18 degrees is 162 degrees. You will bend one half of the roof so it's angled at 18 degrees from the flat work surface.)

Mark the centerline of the roof with a line running parallel to its 48" dimension. Set up a homemade brake (page 37) so the bend will be along the centerline. Bend the sheet slowly and in small increments, stopping frequently to check your angle with the angle guide. Be careful not to exceed the angle.

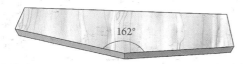

162°

Roof bending guide

14. INSTALL THE ROOF.

Position the roof so it is centered on top of the coop. Use a felt-tip marker to trace the outline of each 4×4 post onto the underside of the roof. Remove the roof.

In the center of each 4×4 outline, drill a ³⁄₁₆" hole.

As discussed earlier, the steel roof will rust if it's left unfinished, so we had ours powder-coated after it was bent and drilled. Alternatively, you can protect your roof with a quality exterior paint or other finish designed for metal (be sure to follow the manufacturer's directions carefully for a long-lasting bond).

Position the roof on top of the coop as before and fasten it to each post with one 1½" roofing screw.

CONTAINER COOP

The Container Coop is meant to be a prototype for a small farm or community garden, with a capacity of up to 20 chickens. The idea for the design came from a request by City Slicker Farms for a coop to use on one of their farms in West Oakland. They gave us a loose program of what they were looking for, with the main criterion being that the coop be ultra-secure from both animals and people. Noting the features of the rest of the site, which included corrugated metal planter boxes and a cedar plank fence, we thought this might be a reasonable time to do something we've always wanted to do: modify a shipping container.

Being near the port, West Oakland has a rich shipping history, with vistas of bayside cranes and expansive stacks of colorful containers lining the water. We went down to the port to a container retailer and picked out a 20-footer, which is half the size of a standard shipping container (see The Shipping Container, on page 184).

City Slicker Farms wanted the structure to meet two needs: a chicken coop with run attached, and a storage shed with an outdoor table to process produce. Once the sections were cut in half and separated, we could start building them out. The extended run for the coop is meant to form the shape of the original container. For the coop and shed siding, we sourced some old redwood planks from a fence that was taken down. The wood siding is meant to offset the harshness of the container. To soften it further, we added sliding barn doors so the shed would read as a farm outbuilding.

Keep in mind that when buying a used container, you have no idea what has been in it. Therefore, we recommend either taking out the existing plywood or hardwood floor or covering it with building wrap and plywood, as we did, to minimize any unwanted exposure for the chickens. You could also buy a 10-foot container and modify it similarly if you want only a coop. The instructions here cover building only the coop half of the container.

The Shipping Container

The shipping container itself was a revolutionary idea, and like many ideas before it, its implementation was slow and cumbersome. Before Malcom McLean lobbied for the standardization of cargo containers to be the same size as a truck's loading space, handling irregularly packed goods was a very profitable business; price was determined by a variety of factors, and cargo was unloaded by small armies of longshoremen who could easily skim off some of the goods coming through the port. By the mid-1960s, aided by a contract with the U.S. government to send supplies to Vietnam, McLean's Sea-Land Services corporation laid the groundwork for a rapid acceleration of global operations, resulting in a subsequent consolidation of labor and a reduction in the number of dockworkers. The overall benefits of streamlining the shipment of goods throughout the world are debatable, particularly if one scrutinizes the type of goods that are shipped thousands of miles using precious nonrenewable resources. For better or worse, however, we are left with a startling number of standardized metal boxes.

Due to the imbalance of goods traded between the United States and China, many containers arrive at American ports full of goods, then start to pile up empty. Because of this surplus, acquiring an empty container to modify is relatively inexpensive. This kink in the global economy has led to a boom in "cargotecture" over the past decade as designers and architects have begun to employ the container as a vessel to be experimented with and expanded upon. There are particular qualities of the container that lend themselves well to such endeavors. As a unit, a shipping container is incredibly structurally sound, made up of load-bearing corrugated steel walls with a supporting frame. A 40-foot container can take a load of up to 30 tons, is sealed from the elements, can be stacked up to four containers high, and has no interior walls or structure to disrupt the volume.

On the other hand, there are significant issues with modifying a container, particularly if its purpose is human (or hen) habitation. The protective paints used to withstand ocean transport have a number of harmful chemicals, and the wood floors used to line the inside of the containers are infused with toxic chemical pesticides to ward off insects. Furthermore, there is no surefire way to determine what contents were shipped in a particular container over its lifetime or what substances may have been spilled in it.

If cut safely and modified properly (by replacing or effectively covering the floor), containers can, however, make for a practical and stylish chicken coop.

MATERIALS

- One-half of a 20-foot shipping container or one 10-foot container
- Four concrete parking bumpers
- 100 square feet waterproof building wrap (such as Tyvek)
- Staples (at least fifty; for building wrap)
- Three 4 × 8-foot sheets ¾" marine plywood
- Thirteen 10-foot 2×4s
- Four 16-foot rot-resistant 4×4 beams
- Two 12-foot rot-resistant 4×4s
- Nine 8-foot rot-resistant 4×4s
- Six 8-foot rot-resistant 2×4s
- Two 10-foot rot-resistant 2×4s
- Eighteen 8-foot pieces 6" rot-resistant siding
- One hundred fifty 1¼" deck screws
- One hundred fifty 3" deck screws
- Thirty 1¼" self-tapping sheet metal screws
- Eight ⅜" × 2½" galvanized bolts with sixteen washers and eight nuts
- Eight 3½" beam hangers
- Twenty-six hurricane tie 4×4 framing connectors with fasteners
- Four exterior 2½" hinges with screws
- One hundred galvanized U nails, 1" length
- Two exterior hasp latches with screws
- Two exterior door handles with screws
- One 48"-long piano hinge
- Two 9"-diameter × 48"-long cardboard concrete tube forms
- 1 bag gravel (for concrete footings; see step 5)
- Four 60-pound bags concrete mix
- Ten 5 × 10-foot sheets galvanized hardware cloth, 14-gauge, with 2" × 2" perforations
- 30 linear feet ½" galvanized hardware cloth, 18-gauge, 36" wide (optional)
- Fifty plastic zip ties

SPECIALTY TOOLS

- Circular saw with wood blade and 7¼" abrasive metal cutoff blades
- Angle grinder with cutoff wheel and sanding disc
- Reciprocating saw with metal-cutting blades
- Staple gun
- Crescent wrench

1. CUT THE CONTAINER.

Prepare for the shipping container delivery by laying out 4×4s on a level area of ground. Have the movers set the container onto the 4×4s to keep it off the ground, to facilitate cutting the container.

Determine where you want to divide the container (we cut ours at an angle) and mark cutting lines along the sides, top, and bottom, using a chalk line. Don safety gear, including a respirator, face shield, ear protection, a leather jacket (or something similar to protect against sparks), and gloves.

Begin cutting the container on the roof, using a circular saw with a 7¼" metal cutoff blade. Change the blade when it gets worn down; you'll have to do this fairly often. Complete the top cut, then cut along each side of the container, starting from the bottom and moving up.

With the side cuts complete, switch to a standard wood blade on the circular saw (or use a reciprocating saw), and set the blade depth to cut through only the wood flooring inside the container. Cut through the flooring along your cut line, starting with a plunge cut (see page 34). Switch back to a metal cutoff blade and cut through the bottom of the container. If you can't reach all of the metal with the circular saw, complete the cut with a reciprocating saw or a large angle grinder.

2. PLACE THE CONTAINER HALVES.

Prepare the site, as needed, for each half of the container, making the ground as level as possible. Position a concrete parking bumper to support each corner of the container half, with the bumpers set at a 45-degree angle (or as desired) to the corners. We added 2×4 bracing in an X across the opening of each container to keep them from buckling while being moved.

Move each container half into place on top of the bumpers. We hired an expert shed mover for this operation; it's possible to do it yourself with the right rigging tools, but only if you're very careful and know what you're doing. One method is to use a couple of heavy-duty jacks and roll the container on round posts.

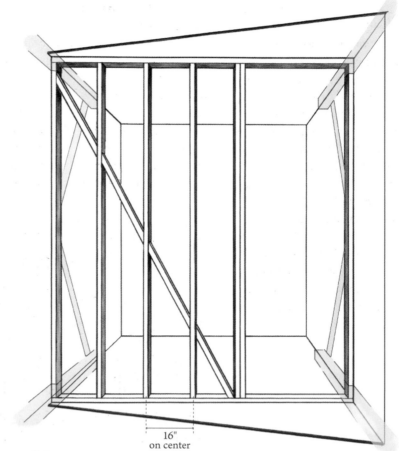

16"
on center

Coop wall framing

3. BUILD THE COOP FLOOR AND WALL FRAME.

Line the inside floor of the coop container with building wrap, stapling it to the existing flooring. Cover the entire floor with a new layer of marine plywood and fasten it to the old flooring with 1¼" screws.

Frame a 2×4 stud wall to enclose the front end of the coop, as shown. This wall runs parallel to the back wall and is bolted to the side walls of the container. Because the corrugated container walls have "hills" (high spots) and "valleys" (low spots), plan the wall so that the two outside studs fall over a hill. You can frame all or part of the wall on the ground first or frame it entirely in place. Depending on your container's construction, you may have to notch the tops of the outside studs to accommodate a steel beam or other structural members. We added diagonal blocking to the stud wall for some extra shear strength to support the container walls.

Space the wall studs 16" on center, and locate the door as desired. When the frame is complete, drill ⅜" holes through the outside studs and container walls, and anchor the studs to the steel walls with four ⅜" machine bolts, washers, and nuts on each stud.

4. INSTALL THE CHICKEN RUN BEAMS.

The frame for the chicken run starts with four 16-foot rot-resistant 4×4 beams that extend horizontally from the front of the coop and are joined to the 2×4 coop wall with notched joints. To install the two bottom beams, cut a 3½"-wide by 1½"-deep notch into one face of each 4×4, starting 44½" from one end. These notches should fit snugly over the outside studs of the 2×4 wall. Make the notches with a circular saw and chisel, as described in Cutting Notches, on page 32.

Set the two bottom beams into position, with the notches fitted over the wall studs. Prop up the free end of each beam, using a stack of scrap wood or a temporary 2×4 brace screwed into the beam, so the beams are level with the floor of the coop. Fasten the beams to the wall studs with several 3" screws. Cut a 2×4 brace to fit at an angle between the rear of the beam and the middle of each outside wall stud and install it with 3" screws.

Measure the distance between the two horizontal beams at the stud wall and cut a 2×4 to that length. Screw the 2×4 near the free (outside) ends of the beams to serve as a temporary support to ensure the beams are parallel to each other.

Notch two more 16-foot 4×4s for the top horizontal beams of the run frame. These notches are the same size as those for the bottom beams but start 41" from the rear ends of the beams (the top beams extend 3½" beyond the bottom beams at the outside end of the run). With help from friends, install the top horizontal beams, using temporary bracing running to the ground to support the free ends. Fasten the beams to the outside wall studs as with the bottom beams. Add 2×4 angle braces at the rear ends of the top beams.

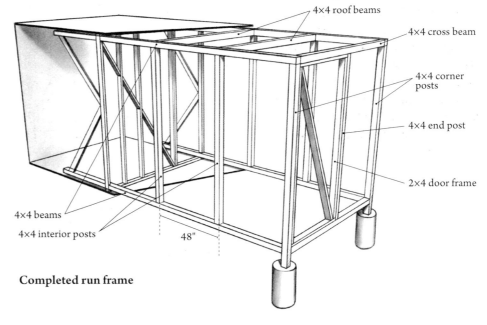

4×4 roof beams

4×4 cross beam

4×4 corner posts

4×4 end post

2×4 door frame

4×4 beams

4×4 interior posts

48"

Completed run frame

5. COMPLETE THE RUN FRAME.

Measure the distance between the top and bottom horizontal beams where they meet the container and cut two 8-foot 4×4 posts to that length. Position the posts vertically between the top and bottom beams on each side of the run so their centers are 48" from the ends of the top horizontal beams. (All of the run's internal posts must be 48" on center so that the mesh pieces will butt up to each other cleanly.) Tack the posts to the beams with 3" screws, then reinforce the joints with metal framing connectors and the manufacturer's recommended fasteners.

Dig a 36"-deep hole for each of the two corner posts, centering the holes under the ends of the bottom beams. Make the holes large enough for a 9"-diameter concrete tube form. Add a few inches of gravel to each hole and tamp it down with a 4×4. Cut each 4×4 corner post (from 12-foot lumber) to length so it stands in the hole and fits snugly against the bottom of the top horizontal beam. Cut each concrete form to length, as needed, and set it in the hole so it extends several inches above ground level. Plumb and backfill around the form. Set the corner posts and make sure they're plumb, then anchor them to the top and bottom beams with beam hangers. Fill the forms with concrete, overfilling them slightly and shaping the top into a dome that will help shed water away from the posts. Let the concrete cure as directed by the manufacturer.

Cut and install the remaining 4×4 framing members, as shown in the *Completed run frame* on page 187. These include two more interior posts on each side of the run (spaced 48" on center); two horizontal crossbeams spanning between the outside ends of long side beams; one post between the crossbeams and centered on the end of the run; and two roof beams located over the side posts. Reinforce all joints with hurricane tie framing connectors, and use beam hangers for the roof beams. Remove the temporary bracing for the long side beams.

Build a door frame out of rot-resistant 2×4s, making it ½" narrower and shorter than the rough opening at the end of the run frame. Include cross-bracing to prevent the door from sagging. Mount the door frame to the corner post with four 2½" hinges.

6. INSTALL THE COOP HARDWARE CLOTH.

Dig a trench along the sides and front of the run area for burying the hardware cloth, which should extend at least 16" into the ground (see page 16). Also dig along the front edge of the coop floor for enclosing the space between the ground and the bottom of the shipping container.

Cut and install hardware cloth on the sides and top of the run framing, aligning the mesh edges with the center of each framing member. Fasten the mesh to the framing with galvanized U nails driven every 12". It's easiest to cut this heavy-gauge mesh with a circular saw with a metal cutoff blade or an angle grinder.

Extend the hardware cloth straight down into the trenches. Tie neighboring sheets together below the bottom beams, using zip ties. As an option, you can also add a layer of ½" × ½" galvanized hardware cloth along the bottom openings of the run to keep out rats and other digging animals and to secure any gaps around the concrete post footings. Backfill the trenches with soil and tamp it thoroughly so the mesh is secure.

7. ADD THE COOP SIDING.

Cover the framed wall of the coop with siding, spacing the boards about ½" apart for ventilation. Fasten the siding with 1¼" screws. Also cover the front side of the chicken run door frame with siding, then install a hasp latch and door handle on the door.

Mesh installed at corner of run

Siding and hardware on chicken run door

Siding on coop wall

8. CREATE THE EGG DOOR AND VENTILATION WINDOW.

Cut a tall, narrow opening into one side of the coop container, using a circular saw with a cutoff blade, as before. The opening on our coop is 84¼" tall and 25" wide. It's best to locate the opening on a flat section of the container, if available.

Build a frame for the egg door with rot-resistant 2×4s, making it at least 2" wider than the cutout opening. Cover the front of the frame with vertical siding boards, butting the boards together side by side. Mount the door to the container wall with a piano hinge and 1¼" self-tapping sheet metal screws.

Build a frame for the ventilation window above the egg door, using rot-resistant 2×4s and mitering the corner joints. This frame should be the same width as the egg door. Cover the back side of the frame with galvanized mesh, then screw the frame to the container wall from the inside, using sheet metal screws or through bolts. Leave a ¼" gap between the door and the window so the door can open and close easily.

Outfit the inside of the coop with nesting boxes and roosting bars, as desired, to complete the coop construction (see pages 15 and 17 for tips on planning and installing nesting boxes and roosting bars).

THE OTHER HALF

What can be done with the other half of the shipping container? We moved ours to the other side of the small farm and built it out as a storage shed. We constructed a 2×4 stud wall across the open face of the shed, similar to the coop's, framing out a double-wide opening. We built two big barn doors and covered them and the stud wall with reclaimed redwood fence material. The doors were installed on a sliding track with barn door hardware, and we welded a locking mechanism out of tube steel. With some leftover lumber from the project, we built a ramp for the wheelbarrow. Along the inside we bolted 2×4s horizontally, for tool hangers and shelving. It is a great shed for a small farm, and the coop and shed complement each other nicely.

Now that you've got the coop all picked out, it's time to read up on raising healthy, happy chickens! Whether you're an absolute beginner and want some heavy hand-holding or you're experienced and want to raise heritage breeds, Storey has the books to help you learn more about raising and keeping chickens.

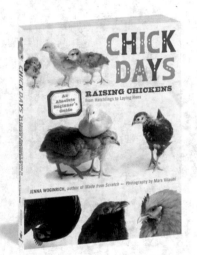

CHICK DAYS
by Jenna Woginrich
Photography by
Mars Vilaubi
A delightful photographic guide for absolute beginners, chronicling the journey of three chickens from newly hatched to full grown.
128 pages. Paper.
ISBN 978-1-60342-584-1.

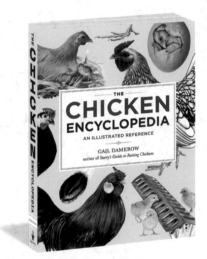

THE CHICKEN ENCYCLOPEDIA
by Gail Damerow
From albumen to zygote, the terminology of everything chicken demystified.
320 pages. Paper.
ISBN 978-1-60342-561-2.

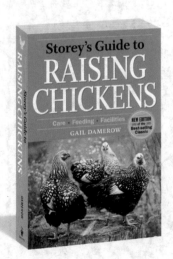

STOREY'S GUIDE TO RAISING CHICKENS
by Gail Damerow
The ultimate guide that includes information on training, hobby farming, fowl first aid, and more.
448 pages. Paper.
ISBN 978-1-60342-469-1.
Hardcover.
ISBN 978-1-60342-470-7.

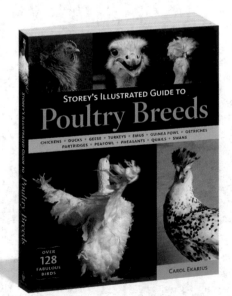

STOREY'S ILLUSTRATED GUIDE TO POULTRY BREEDS
by Carol Ekarius
A definitive presentation of more than 120 barnyard fowl, complete with full-color photographs and detailed descriptions.
288 pages. Paper.
ISBN 978-1-58017-667-5.